Street Sounds

Street Sounds
Listening to Everyday Life in Modern Egypt

Ziad Fahmy

Stanford University Press
Stanford, California

STANFORD UNIVERSITY PRESS
Stanford, California

© 2020 by the Board of Trustees of the Leland Stanford Junior University.
All rights reserved.

No part of this book may be reproduced or transmitted in any form or by any means, electronic or mechanical, including photocopying and recording, or in any information storage or retrieval system without the prior written permission of Stanford University Press.

The book epigraph is from *Listening to Nineteenth-Century America* by Mark M. Smith. Copyright © 2001 by the University of North Carolina Press. Used by permission of the publisher. www.uncpress.org

Printed in the United States of America on acid-free, archival-quality paper

Library of Congress Cataloging-in-Publication Data is available on request.

LCCN 2020937854 | ISBN 9781503612013 (cloth)| ISBN 9781503613034 (paper)

Cover design: Kevin Barrett Kane

Cover photograph: Egypt, car traffic waiting for crossing pedestrians in Cairo, 1961. © Harrison Forman. American Geographical Society Library, University of Wisconsin-Milwaukee Libraries.

Text design: Kevin Barrett Kane

Typeset at Stanford University Press in 10/15 Minion Pro

This book is dedicated to the memory of my brother,
Tarek Fahmy (1974–2014), and of my father,
Adel Muhammad Abd al-Aziz Fahmy (1946–2014).
They both still live in my thoughts, and in many ways,
they remain a part of who I am as a person.

It seems almost audacious to point out that in the past, peoples sensed their worlds, their environments, and their places and mediated their experiences sensorially. Obvious though this fact is, however, it warrants stating not least because we are prone to examine the past through the eyes of those who experienced it. While people interpreted their world visually, it is also worth iterating that seeing was but one way in which they experienced.

Mark M. Smith, *Listening to Nineteenth-Century America*

Contents

Maps, Figures, and Tables xi

Note on Transliteration and Translation xv

Preface xvii

Acknowledgments xix

INTRODUCTION
    Historicizing Sounds and Soundscapes 1

Part I Mundane Street Sounds and Ownership of Public Space 23

  1 Walking the City: Street Voices, Traffic, and the Mundane Sounds of Everyday Life 25

  2 Silencing the Streets: Classism, Fear of the Crowd, and Regulating Sounds and Bodies 53

Part II Infrastructure, Technology, and the Sounds of Modernity 81

  3 Roads and Tracks: Modern Traffic and the Sensory and Social Impact of Trams and Automobiles 83

  4 The Soundscapes of Modernity: Electricity, Lights, and the Sounds of Nightlife 120

Part III The Sounds of Public Spectacles: Between Ordinary People and State Legitimacy 155

  5 The Sounds of Weddings and Funerals: From Brass Bands to Wails and Ululations 157

  6 Sounding Out State Power: Cannons, Music, and Loudspeakers 189

CONCLUSION
    Class Distinction and Remembering Lost Sounds 215

Notes 227

Bibliography 261

Index 281

Maps, Figures, and Tables

Maps

MAP 3.1	Cairo's Roads and Squares	87
MAP 3.2	Cairo's Tramway and Bus Routes (1944)	104
MAP 4.1	Azbakiyya and Cairo's Entertainment District	139

Figures

FIGURE 1.1	Street Microphone	28
FIGURE 1.2	Hand on Car Horn	29
FIGURE 1.3	Cart Driver and His Assistant	30
FIGURE 1.4	Guava Seller	37
FIGURE 1.5	Cairo: Women Drawing Water from the Nile	43
FIGURE 1.6	Licorice Drink Seller and his Brass Saucers	47
FIGURE 2.1	Confrontation between Police and Street Merchants in Port Said (1900)	62
FIGURE 2.2	Simit Bread Seller (ca. 1955)	64
FIGURE 2.3	Egyptian Peasants Caricatured as Chimpanzees	67
FIGURE 2.4	"Is He Blind or Is He Pretending to Be Blind?"	72
FIGURE 2.5	A Gang of Young Boys	76
FIGURE 3.1	Mishmish Afandi and Traffic Noise	106
FIGURE 3.2	Al-Ataba al-Khadra Square in 1907	109
FIGURE 3.3	Al-Ataba al-Khadra Square in 1947	110
FIGURE 3.4	Station Square: Outside Cairo Railway Station (1900)	111
FIGURE 3.5	Station Square (ca. 1950)	112
FIGURE 3.6	Crowded Cairene Open Tram (ca. 1961)	115

xii   Maps, Figures, and Tables

FIGURE 4.1 Two Men Playing Backgammon in a Cairene Café (ca. 1961)   131
FIGURE 4.2 Outdoor Café with Gas Streetlights (ca. 1900)   133
FIGURE 4.3 Radio Noise in the Streets   137
FIGURE 4.4 Kursaal Nightclub on Imad al-Din Street (1916)   138
FIGURE 4.5 Eldorado Café (ca. 1908)   141
FIGURE 4.6 Neon Signs and Cairo Lights (ca. 1950)   146
FIGURE 4.7 Movie Theater and Café on Imad al-Din Street (1925)   149
FIGURE 4.8 The Strand Cinema in Raml Station Square in Alexandria (ca. 1939)   151
FIGURE 5.1 Egyptian Army Band Returning to Barracks (1907)   164
FIGURE 5.2 Cairene Marriage Procession (1898)   169
FIGURE 5.3 A Wedding Procession (ca. 1910)   171
FIGURE 5.4 Funeral Procession Nearing the Cairo Cemetery (ca. 1910)   179
FIGURE 5.5 Two Professional Male Mourners (ca. 1900)   183
FIGURE 6.1 The Funerary Parade of Khedive Ismail at Opera Square (March 22, 1895)   192
FIGURE 6.2 Egyptian Army Cannon Used to Announce the End of Ramadan (Eid al-Fitr)   197
FIGURE 6.3 The King's Band   200
FIGURE 6.4 "At the Mawlid"   203
FIGURE 6.5 Mawlid al-Nabi in Cairo's Abbasiyya Grounds: The Awqaf Ministry Pavilion (1914)   206
FIGURE 6.6 Military Band at the Prophet's Birthday Parade in Abbasiyya (January 1949)   208
FIGURE 6.7 The King Recording His First Radio Speech (May 8, 1936)   210
FIGURE 6.8 Loudspeakers in Cairo's Streets   211
FIGURE 6.9 Nahhas Pasha Reading the King's Speech in Parliament   212

Tables

TABLE 3.1 Electric Tramways in Cairo and Alexandria (1899–1950)   93
TABLE 3.2 Licensed Animal-Drawn Vehicles in Egypt (1891–1953)   95
TABLE 3.3 Registered Motorized Vehicles in Cairo and Alexandria (1904–1921)   98

TABLE 3.4	Licensed Motorized Vehicles in Egypt (1925–1954)	102
TABLE 4.1	Yearly Imports and Costs of Electrical Appliances in Egypt (1934–1950)	128
TABLE 4.2	Caféss, Bars, Restaurants, and Nightclubs in Egypt (1937–1947)	130
TABLE 4.3	Cairo's Major Cabarets and Theaters (ca. 1950)	147
TABLE 4.4	Cairo's Major Movie Theaters (ca. 1950)	148

Note on Transliteration and Translation

Fusha (pronounced "fuss-ha"), or classical Arabic, has been transcribed according to a simplified system based on the translation and transliteration guide of the *International Journal of Middle East Studies* (*IJMES*). All diacritical marks have been omitted except for the '*ayn* ('), and the *hamza* ('). For texts, songs, and plays written or performed in colloquial Egyptian, I have slightly modified this system. Instead of *jim* (j), I use *gim* (g); instead of *qaf* (q), I use a *hamza* ('). Also, the definite article is transliterated from *fusha* as *al-*, and from colloquial Egyptian as *il-*. The names of Egyptian authors writing in French or English have not been changed.

Unless otherwise noted, all translations from Arabic are mine.

Preface

Growing up in Alexandria, Egypt, from the mid-1970s to the mid-1980s, I was surrounded by some unique sights and sounds. Our balcony in Rushdi Street (near Stanley Bay) was a place of sensory wonder for my brother and me, as we listened, observed, and sensed our immediate neighborhood from that third-floor perch. Among the repeating sounds we heard on most mornings were the distinctive and loud rhythmic thumps produced by the seemingly ritualized beatings of Persian carpets from neighboring balconies and windows. The force of the specialized wicker carpet-beaters produced not only these booming bangs but also airborne clouds visible for a few seconds, layered with the odor of dust that for a few minutes added a third sensory dimension to the neighborhood's sights and sounds. With vacuum cleaners becoming more prevalent in the 1980s, carpet beating began to decrease in upper and middle-class households, and soon the practice will altogether disappear. Other neighborhood sounds included the loud cry of "*bikyya*" from the *robabikya* junk collector,[1] and the unintelligible holler of the fava bean seller, who every morning stopped by our building with his wooden cart. I found out years later that he was yelling out "*luz ya-ful*" (beans like almonds), implying that his fava beans tasted and looked like almonds. Until the early 1980s, the milkman arrived on his bicycle, ringing his bell to alert the neighborhood. He had two large stainless-steel containers balanced on each side of his bicycle. Today, the milkman and fava bean seller do not make their rounds in Rushdi and other middle- and upper-middle-class urban areas, yet the still ubiquitous sounds of robabikya junk collectors remain.

When, at the age of eleven, I migrated with my family to the suburbs of central New Jersey, street sounds remained an important part of my new sensory space. In East Brunswick, the *bikyya* calls were replaced with the loud bells of the Good Humor ice cream truck and the jingles playing from the loudspeaker of the Italian ice truck. Some of the natural sounds characterizing both places were also significantly different. For instance, the loudness and the intensity of summer thunderstorms in central Jersey were measurably more deafening than Alexandria's seasonal rainstorms.

As an eleven-year-old, experiencing my first central Jersey thunderstorm with the frequent sonic booms of nearby lightning strikes was downright terrifying. Heavy rains and stormy winds arise in Alexandria only in the winter, thunder is rare, and lightning strikes are unheard of. Conversely, snowy winter nights in central Jersey were noticeably quieter, as the blanketing layer of snow muffled surrounding sound and made for eerily silent nights. Thinking back, I am amazed by how much these and other natural and manmade sounds were integral to my sensory and intimate comprehension of both spaces. In fact, as I elaborate in this book, sound can reveal a great deal about a place to both its residents and to historians examining the past.

Working toward this end, *Street Sounds* examines how everyday people dealt and engaged with their sonic environment. I will be examining sounds and sounded phenomena, and I will also be using sounded sources as a key analytical tool for examining Egyptian street life, especially accounting for the dramatic sonic changes resulting from the successive introduction of modern transportation, lighting, and amplification technologies. Ordinary Egyptians sounded out their presence not just through publicly marketing their wares and loudly celebrating or mourning life's milestones and tribulations but also through their everyday banal movements, their daily commutes, and their mere existence in the streets and public squares. By listening in to the voices of ordinary people and by documenting how they physically and sonically occupied, used, and misused the streets, we can begin to uncover more of their agency—despite the state's constant interference—in appropriating the streets as their own. This book is also about the politics of sound, and sound's entanglement with issues of class formation and changing conceptions of modernity and identity. *Street Sounds* addresses the sensory politics of sound and "noise" by demonstrating how the growing middle classes set out to sensorially distinguish themselves from the masses. Finally, the dimension of sound not only contextualizes space by adding to it another layer of texture and meaning but also brings us closer to the streets and street life as lived and experienced by everyday people.

Acknowledgments

The initial idea for *Street Sounds* came about in early 2011, as I was finishing the final revisions of my previous book, *Ordinary Egyptians*. Because I was dealing primarily with recorded music, vernacular theater, *zajal* (colloquial poetry), and other aural and oral sources for that book, I had become more consciously aware of the importance of sound and listening to not only the project at hand but more broadly to history writing in general. It became obvious to me at that time that more sounded histories of Egypt and the Middle East were sorely needed.

Financial assistance for the research and writing of this project came from a variety of sources. In the summers of 2012 and 2013, two generous seed grants from Cornell's Mario Einaudi Center for International Studies and the Institute for the Social Sciences (ISS) Research, allowed me to spend four months in Kew, London, to conduct research at the British National Archives. The bulk of the research for this book took place in Egypt in 2013 and 2014, with the help of generous research grants from a National Endowment for the Humanities (FPIRI Program)–American Research Center in Egypt (ARCE) Faculty Research Fellowship. The NEH-ARCE fellowship allowed me to spend ten months in Cairo, where I conducted research at the Egyptian Archives (Dar al-Watha'iq) and the Egyptian National Library (Dar al-Kutub). Throughout the writing process, the academic staff in the Department of Near Eastern Studies at Cornell University were generous and supportive as I juggled my writing with my chairing responsibilities. Thank you Julie Graham, Christiane Capalongo, and Ayla Cline. I am also indebted to all of my colleagues in the Department of Near Eastern Studies for their intellectual and moral support.

I also thank Ali Houissa, Middle East and Islamic Studies Librarian at Cornell University, for acquiring dozens of rare Arabic periodicals, which were extremely useful for my research, and also Susan Peschel, Visual Resources Librarian for the American Geographical Society Library at the University of Wisconsin Milwaukee, who was most helpful in providing some of the digitized images for this book.

Street Sounds took over seven years to research and write and has passed through several stages of transformation. During that time, I have benefited from the encouragement and ideas of countless colleagues, mentors, friends, and family members. My thirst for understanding the importance of soundscapes and listening was happily quenched during my residence as a Fellow at Cornell's Society for the Humanities, where—conveniently for me—the focal theme for the 2011–2012 academic year was "Sound: Culture, Theory, Practice, Politics." During my year at the Society of Humanities, my readings, weekly seminars, and almost daily discussions with the other Fellows from various academic disciplines were instrumental in forging my ideas about sounds and soundscapes. I thank, the society's former director Tim Murray, administrative assistant Mary Ahl, and events coordinator Emily Parsons for making that year such a productive, intellectually stimulating period for me and the other Fellows. For stimulating and constructive conversations and critiques, I thank all of the 2011–2012 fellows, including Eliot Bates, Marcus Boon, Miloje Despic, Sarah Ensor, Andrea Hammer, Brian Hanrahan, Bart Huelsenbeck, Michael Jonik, Damien Keane, Nicholas Knouf, Yongwoo Lee, Eric Lott, Roger Mosely, James Nisbet, and Trevor Pinch. A special shout-out to Duane Corpis, Nina Eidsheim, Jeanette Jouili, and Jennifer Stoever; their unpretentious intellectual generosity and their contributions to our writing workshop made my stint as a society Fellow especially fruitful and enjoyable.

In the last few years, several sound- and sensory-related workshops and symposiums were vital in helping me formulate my ideas for *Street Sounds*. I would like to thank Priyasha Mukhopadhyay and Peter McMurray for organizing the "Acoustics of Empire" workshop held at Harvard University and the subsequent "Acoustics of Empire" conference at Cambridge University, in March and December of 2018. At these two meetings, I benefited from my conversations and interactions with Peter Asimov, Arthur Asseraf, Elleke Boehmer, Alejandra Bronfman, Hyun Kyong Chang, Nick Cook, James Davies, Faisal Devji, Emily Dolan, Katharine Ellis, Edward Gillin, Glenda Goodman, Alexandra Hui, Melle Kromhout, Veronika Lorenser, Nazan Maksudyan, Adam Mestyan, Jairo Moreno, Rumya Putcha, Sindhumathi Revuluri, Gavin Steingo, Jim Sykes, David Trippett, Benjamin Walton, Amanda Weidman, and Richard Williams.

I also thank Shayna Silverstein for organizing the symposium *Listening In: Sonic Interventions in the Middle East and North Africa*, which was held at the Department of Performance Studies at Northwestern University in May 2016. At this symposium, I learned a great deal from Cristina Moreno Almeida, Rüstem Ertuğ Altınay, Sascha Crasnow, Deborah Kapchan, Maria Malmström, Peter McMurray, Wendy Pearlman, Darci Sprengel, Leila Tayeb, and Michelle Weitzel. At the October 2015 workshop "Corporeality in Arab Public Culture," held at the Netherlands Institute of Advanced Studies (NIAS), I exchanged ideas about embodiment in the public sphere with the other participants, including Ines Braune, Farha Ghannam, Marwan Kraidy, Michiel Leezenberg, Vivienne Matthies-Boon, Jared McCormick, Annelies Moors, Sara Mourad, Judith Naeff, Charlotte Pardey, Thomas Poell, Petra Sijpesteijn, and M. R. Westmoreland. Finally, I thank all the participants at the "Music and the Public Sphere" workshop, held in April 2014 at the Berlin Institute for Advanced Studies (WICO). They include Kelly Askew, Thomas Christensen, Jocelyn Chu, Amlan Dasgupta, Eric Drott, Julian Henriques, Kaiwan Mehta, Jonathan Neufeld, Tejaswini Niranjana, Jann Pasler, and Surabhi Sharma.

At the annual Middle East Studies Association (MESA) conferences, several panels and roundtables about sounds and soundscapes have been held in the last few years, culminating in the publishing of a roundtable in the *International Journal of Middle East Studies* in 2016.

Participating in these events, I learned a great deal from Elvan Cobb, Nahid Siamdoust, Andrea Stanton, Carole Woodall, and many others. The Central New York Humanities Corridor Fellowship workshops that I attended at Cornell University and Syracuse University with Carol Fadda-Conrey, Timur Hammond, Amy Kallander, Mostafa Minawi, Kent Schull, Nazanin Shahrokni, and Mary Youssef were a very productive space for encouragement and incisive critiques. Closer to home, in the spring of 2019, I presented selections from *Street Sounds* at the Cornell History Department's Comparative History Colloquium. I thank Ibrahim Gemeah, Durba Ghosh, Sandra Greene, Amr Leheta, Mostafa Minawi, and the other colloquium participants for critically commenting on my Introduction and Conclusion.

Yet again, I thank Kate Wahl, Stanford University Press Publishing Director and Editor-in-Chief, and her great editors and staff, including Faith Wilson Stein, Gigi Mark, and Elspeth MacHattie, for taking a chance on a history book about sounds and soundscapes. Kate's thoroughness and professionalism made the entire process smooth, flawless, and stress free. The critical and encouraging comments of the SUP anonymous readers were extremely helpful in framing my revisions. My book is much better because of their comments and advice. I am especially indebted to Joel Gordon and Walter Armbrust for their critical and detailed chapter-by-chapter comments on my entire book. Their insights led me to explore new paths I had not yet considered.

I offer a special thank you to Orit Bashkin, Aomar Boum, James Gelvin, Kim Haines-Eitzen, Zachary Lockman, Andrew Simon, and Andrea Stanton for meticulously reading and commenting on an earlier version of my manuscript. I also thank Rama Alhabian, Kyle Anderson, and Steve Weed for their critical reading of a recent draft of my Introduction.

Throughout my career, a great number of people have contributed to my overall intellectual development. I am grateful to many colleagues, friends, former professors, advisors, and family members. I thank Israel Gershoni for his advice and overall support of my scholarship. I will forever be in the debt of Charles D. Smith, Julia Clancy-Smith, and Linda T. Darling, whose intellectual insights, patience, and guidance were instrumental in giving me the research and intellectual tools necessary to become a historian. I thank my amazing aunt and uncle, Zahia Fahmy and Kamal El-Shayeb, for hosting me on multiple occasions at my favorite "writing retreat" in Naxos, Greece. The food and wine and the sounds of the wind and sea are always an inspiration. I especially thank Anndrea Mathers, who read and proofread through most of the manuscript and endured my endless obsessions with this book project. Her editorial, intellectual, and emotional support made this book possible.

Most of all I am grateful for my mother, Ferial Sammakia, who is supportive of all of my endeavors. Without her unconditional love and encouragement, I would not be the person I am today.

Street Sounds

Introduction
Historicizing Sounds and Soundscapes

IN LATE FEBRUARY 1936, a correspondent for *al-Radiu al-Misri* (Egyptian Radio) magazine wrote a detailed article describing the 1936 Cairo Agricultural and Industrial Exhibition. Amplifying the goals of the exhibition organizers, the article mainly promoted Egypt's national industrial and economic potential, while also attentively describing the diversity of sounds at the exhibition:

> I sat down in a nice café in front of the Cotton Museum observing the visitors [to the exhibition] as they came and went. It was very crowded and full of people of all social classes, democratically intermingling without a fuss.... As I was sitting alone, I listened carefully to the cacophony that was broadcast from the loudspeaker installed at the top of the Cotton Museum. The announcer read out many commercial advertisements praising the quality of various goods. Afterward, he repeated that the Cairo Exhibition's radio station was sponsored by the marketing offices of various Egyptian corporations and was operated by the [Egyptian] Telephone Company. The station then broadcast some musical recordings and comedic dialogues.... The cacophony produced by the loudspeaker was continuous as intermittently the exhibition's small train blew its loud whistle. All of these various noises were mixed in with the sounds of one of the military brass bands. Adding to the din—and complementing all of these diverse sounds—was the constant and tedious background drone of the steam irrigation pump which was continuously running at the exhibition's agricultural machines department. This drone was akin to a primary tune orchestrating all of these diverse sonic elements, as they simultaneously reached my ears and combined into one composition. All of these sounds were intermixed with the ever-present noise of people's chatter and loud voices. Yes, the clamor was great![1]

The 1936 Cairo Agricultural and Industrial Exhibition was open for two months—from February 15 to April 15—and in this short time, 1.5 million visitors came through its gates.² To put this figure in perspective, in 1937, the entire population of Cairo was around 1.3 million.³ The exhibition was held at the Cairo Exhibition fairgrounds at the southern tip of the Island of Gezira (Zamalek).⁴ Unlike the orientalist representations of Egypt featured at the 1889 World Exhibition in Paris, the 1936 Cairo Exhibition was purposely created to visually and sonically depict Egypt as part of the modern world.⁵

Nineteen thirty-six was an eventful year for Egypt. On April 29, 1936, just a couple of months after the opening of the Cairo Exhibition, King Fuad (r. 1917–1936) died and the young and relatively unprepared King Faruq (r. 1936–1952) assumed the throne of Egypt. Just as importantly, in late August of the same year, the 1936 Anglo-Egyptian Treaty was signed, renegotiating Britain's 1922 unilateral declaration of Egyptian "independence" by giving Egypt more political autonomy. The British military occupation, however, which had started in 1882, would continue until 1956. As can be gleaned from the tone of the *al-Radiu al-Misri* article and from the extensive press and media coverage, the 1936 exhibition was a source of pride for Egyptian nationalists and modernists at a critical juncture in Egypt's road to political and economic independence. To be sure, though, like most exhibitions, the Cairo Exhibition also exemplified commodity fetishism and was built in order to support Egypt's growing capitalists and not just to demonstrate the country's aspirational economic nationalism.⁶

Although exhibitions are often used to theorize about the optical detachment of the visual and the modern, as the quoted commentary demonstrates, the sounds of modernity were just as important and as prevalent as any visual representation.⁷ The blaring loudspeakers, the exhibition's miniature train (used to transport visitors throughout the expansive exhibition grounds), the drone of the motorized water pump at the agricultural exhibit, and the sounds of the military brass bands provided a constant soundtrack to the visual displays of the buildings, industrial and agricultural machines, tractors, automobiles, and a nighttime array of dazzling electric lights. Expanding the sonic reach of the exhibition beyond the fairgrounds, the Egyptian government radio station broadcast the entire opening ceremony of the exhibition to tens of thousands of listeners.⁸ In addition, the exhibition had its very own "radio station" broadcasting locally through loudspeakers placed strategically throughout the grounds. The studio used for these local broadcasts was itself an exhibit, a functioning miniature replica of a radio studio. The exhibit's radio announcer continually played music, read out commercial announcements advertising the various products that were sold or displayed at the exhibition, and occasionally announced the names of lost children, to help reunite them with their parents.

Lest we overlook the other senses, the entire experience of going to the exhibition was multisensory, as the visiting men, women, and children were sensorially immersed in the experience of walking through the exhibits by observing and listening. Most could also smell the burning coal and gasoline fueling the train, tractors, automobiles, water pumps, and other machinery. Visitors no doubt also touched, smelled, and tasted some of the foods and drinks in the many cafés set up within the exhibition grounds. Handling and touching the souvenirs, fabrics, textiles, and other products on display in the many stalls and shops was another integral part of the experience. Although in many ways the sponsors built the exhibition to be an aspirational microcosm representing the future of Egyptian agricultural and industrial modernity, to the majority of the visitors, it was simply a place for family outings and meant strictly for entertainment.

Large crowds of Egyptians of all classes attended the exhibition, including many children, who were especially making use of the branch of Luna Park that was set up especially for the occasion. The elaborate amusement park included a haunted house, roller coasters, various rides, and even bumper cars that, observers noted, were regularly used by children as well as adults. For better or worse, the 1936 Cairo Exhibition was a carnival-like ode to modernity and the potential of Egyptian economic independence. It was a loud and cacophonous affair with loudspeakers playing recorded music, and various traditional and modern brass bands performing live at different venues.[9] Listening to the exhibition, instead of just noting its visual representations, reveals a great deal more about what happened at the ground level among the thousands of ordinary visitors who were strolling about, talking, eating, drinking, and riding the exhibition train or the various amusement park rides.

. . .

In this book, I examine everyday life in Egypt using sound and the politics of sound as one of the key tools for uncovering the changes that went on in Egyptian urban streets during the rapidly shifting first half of the twentieth century. By listening in to the changes materializing in the Egyptian streets, we can get a lot closer to the embodied mundane realities of pedestrians, street peddlers, and commuters. This allows for a more micro-historical examination of everyday people's interactions with each other and helps us evaluate the impact of the various street-level technological and infrastructural manifestations of modern Egyptian life.[10] As the twentieth century roared on, unfamiliar unmediated and mediated sounds were introduced almost year after year, with new technological innovations drastically changing the soundscapes of the streets. These generally loud and transformative inventions, ranging from trains, trams, and automobiles to water pumps, radios, telephones, and loudspeakers fundamentally affected and altered not only the Egyptian soundscape but also the lived public culture of all Egyptians.[11] Indeed, from the last quarter of the nineteenth century to the middle of the twentieth century, the steady introduction of these new and unfamiliar sounds not only added to the soundscape but, by gradually drowning out and at times intermingling with other quieter manmade and natural sounds, also modified some of the more traditional sounds of everyday life.[12] The growling of automobile and motorcycle engines and the hum of fluorescent lights and later on radios, refrigerators, fans, and air conditioning masked and concealed as much noise as they produced. In an urban environment, one was more likely to hear footsteps, street-side conversations, the rustling of leaves, the wind, and birds and other animals in the late nineteenth century than in the 1950s. Today, it can be difficult to imagine how a town or a city sounded in the late nineteenth century, though listening carefully during a major power outage can reveal somewhat the degree, volume, and variety of "noise" that our plethora of electrical appliances and devices produces and can also remind us of the sounds this machinery conceals.

It is impossible to overestimate the role that electricity played in completely transforming twentieth-century society. The gradual and uneven introduction of electricity in Egypt, dramatically and forever changed most aspects of Egyptian everyday life, especially changing what people saw and heard, indoors and out.[13] Telephones, radios, electric microphones, and electric recording and amplification technologies transformed how people received and processed information, misinformation, gossip, rumors, propaganda, and entertainment. And it was not just these audible devices that had an impact on the urban soundscape. In the early nineteenth century, for example, Cairo's many quarters would literally shut down their large wooden doors at night, as darkness and relative silence enveloped most of the city. Municipal gas lamps and later on electric lighting forever changed the sounds of the night. Egyptians would more regularly stay up later at night than ever before, whether by visiting well-lit cafés, theaters, cinemas, amusement parks, stores, and markets, or by staying at home in an electrically lit dwelling. A regular "everyday" nightlife, with all of its entertainment, leisure, commercial, and sonic implications, was only possible with the spread of electricity and electric lights.

Street Sounds is the first historical examination of the changing soundscapes of modern Egypt. In the following pages, this book documents the street-level effects of this sonic transition, not to examine these sounds for their own sake, but to understand the wider cultural and class implications of this sounded technological transformation and to assess its impact on Egyptian street life. The book tunes into the sounds of the past through a careful analysis of historical texts in order to assess the street-level, evolutionary impact of aural modernity. *Street Sounds* also addresses the sensory politics of sound and "noise," and critically examines the intersection of state power with street life as the state attempted to control the streets. Just as importantly, it accounts for the growing middle classes as they set out to sensorially distinguish themselves from the Egyptian masses. By considering the changing sounds of modern Egypt, this book not only accounts for the large-scale urbanization and modernization rapidly taking place but, more importantly, it also amplifies some of the voices and noises of those who actively participated in this ever-changing sonic environment. Beyond examining sounds and sounded phenomena, I will be using sounded sources as one of my key analytical tools for investigating Egyptian street life, and especially for analyzing the dramatic sonic changes resulting from the successive introduction of modern transportation, lighting, and amplification technologies. Finally, *Street Sounds* proposes that by taking into account the changing sounds of the past, and by examining how people dealt with their daily sonic environment, a closer, more embodied, microlevel analysis of everyday life is possible.

Historians and Soundscapes

Historians have recently started listening to the past, contributing to what David Howes has described as a "sensorial revolution in the humanities and social sciences."[14] Mark Smith, one of the history of sound pioneers, has triumphantly declared that historians today are "listening to the past with an intensity, frequency, keenness, and acuity unprecedented in scope and magnitude. Once focused on just the history of music and musicology, historians of aurality now consider sound in all its variety."[15] With a few recent exceptions, historians of the Middle East have yet to explore that path, thus unintentionally portraying the past as silent, and devocalized.[16]

In order to pursue a more sounded approach to history, historians can learn from other disciplines that have already made progress in integrating auditory techniques. Anthropologists along with media studies scholars are leading the way in a recent explosion of media studies focused on the contemporary Middle East. Dozens of works covering contemporary satellite television, movies, and music and sound recordings have been published in the first two decades of the twenty-first century.[17] Though most of these works do not address aurality and sound directly, by virtue of their subject matter, they all offer scholarly examinations of sounded sources. Works on various forms of Arab media by Charles Hirschkind, Ted Swedenburg, Lila Abu Lughod, Marwan Kraidy, Flag Miller, and Walter Armbrust are conceptually useful for historians. Hirschkind's *The Ethical Soundscape* in particular—which engages with and is inspired by the burgeoning literature in sensory and aural studies—is a good starting point for much needed future studies on contemporary and historical Middle Eastern soundscapes.[18] To be sure, most anthropologists and media studies scholars deal with contemporary societies, and hence are able to directly listen to the sounds and soundscapes they are studying. Historians do not have that luxury: we must use different strategies to cull and tease out sound from mostly textual sources. In addition, most of these studies deal with music, sermons, television programs, and film and not with noise, street sounds, and soundscapes.

There are many more historical dimensions to be discovered if we are open to considering sound as a serious path of inquiry for understanding the past. Sound historian Jonathan Sterne has accurately declared that "there is always more than one map for a territory, and sound provides a particular path through history."[19] More specifically, by taking into account the changing sounds of the past, and by considering how everyday people dealt with their daily sounded environment, we can provide valuable texture and context to historical examinations of everyday life and be brought closer to a more embodied microlevel analysis of street life. This is especially true in a period of rapid sonic transition, such as the infrastructural and technological transformations occurring at the turn of the twentieth century. Listening in to the sources brings us closer to the lived contemporaneity of past experience, revealing in the process how hearing and sound were critical to understanding everyday life.

In *Street Sounds*, I examine some of these dramatic infrastructural and technologically driven sonic changes, while also keeping tabs on the ongoing cultural and mass media representations and classifications of soundscapes, sound, and noise. All the while, I will be closely examining the developing discourses on class and modernity in relation to the changing soundscapes and the evolving construction and reconstruction of a modern Egyptian notion of noise. By "listening" to the changing sounds of early- to mid-twentieth-century Egypt, I hope not only to take into account the large-scale urbanization and modernization rapidly taking place but, more importantly, also to historicize and inquire into the significance of some of the voices and noises of those who actively participated in this ever-changing soundscape. As I demonstrate in the pages ahead, there is much to uncover if we incorporate sounds and soundscapes as part of our methodological toolkit for understanding the past. In part, *Street Sounds* will trace how sound and hearing fit within the broader framework of encountering, appropriating, and resisting street-level twentieth-century Egyptian modernity.

Class, Noise, and Silencing the Masses

The selective weaponization of discourses about loud "lower-class" traditional noises as a tool of social distinction and classism is one of the repeating themes of this book. To paraphrase Jennifer Stoever, there was a clear sonic class line expressed in elite and middle-class discourses about Egypt and the Egyptian streets. The first few decades of the twentieth century were marked by increasingly modernist bourgeois sensibilities, calling for quieter streets and advocating for silencing traditional working-class street sounds.[20] Egypt's growing middle classes played a key role in the way that Egyptian cultural production not only ranked the senses but also attempted to aesthetically define what was considered to be civilized and "in good taste." Judgmental dualities concerning a changing definition of taste, class, and modernity infused most discussions about sounds and noise. How a person smelled; their grooming and clothing habits; the loudness of their voice; their tastes in music, "culture," and food; and how they worshiped, danced, celebrated, mourned, ate, and drank were subject to critique as part of evolving definitions of what constituted "civilized" and "proper" behavior. As was to be expected, the vast majority of Egyptians fell outside these exacting parameters and were castigated and marginalized in the press and in many of Egypt's mainstream cultural productions. Indeed, upon closer inspection, the formation of the middle class and its differentiation from the "traditional" Egyptian masses partly created the need for these discourses, which permeated the Egyptian media in the first half of the twentieth century.[21]

Most of Egypt's print productions in particular, from newspaper editorials to novels, textbooks, and a variety of nonfiction books, advocated for the de facto civilizational uplift of everyday Egyptians and were often judgmental about sensory issues. Class, race, and gender were also associated with certain sounds, accents, and the expected volume of conversational interactions. The literature of the time is replete with modernizing Egyptians preaching to the masses about how loud and vulgar they were and the need for more "civilized" silence. Surprisingly, these discourses of sensory shaming came from an ideologically diverse crowd. Islamists advocated for a demure and homogeneous orthodoxy, as opposed to a more raucously diverse popular Islam, and the modernizing secular elite idealized and emulated an imagined somber and allegedly quiet European public demeanor.[22] If one were to take some of the sensory discourses in the twentieth-century Egyptian press at face value, one could readily find new norms and ideals for how a "modern" citizen should behave, smell, sound, and look in public. Class differentiation and distinction usually relied upon sensory markers, especially auditory ones, separating the cultural elites and the upwardly mobile middle classes from the "loud and unwashed" masses. As this book will show, most discussion about sounds and noise were infused with class anxieties and judgmental dualities concerning changing definitions of taste, class, and modernity.

Ironically, "modern" and educated Egyptians were supposed to be quiet, yet "modernity" in the form of the ever increasing numbers of new machines and inventions introduced in Egypt from the mid-nineteenth to the mid-twentieth century was unabashedly loud. Whereas new industrial and machine sounds were, early on at least, positively associated with modernity and economic advancement, sounds and voices projected by the masses were vulgarized as noise, at best a nuisance and at worst a threat to the established order. This brings us to a rather important question. What is *noise*? The simple definition of noise is unwanted sound. Yet, unwanted by whom? Again, we find that there were often class implications in which sounds were classified as noise and which were considered pleasing or at the very least neutral and non-offending.[23] The many public discourses about sounds and noises that we will consider are full of class and classist implications, with varying attitudes, discussions, and negotiations about class and modernity. These were not so much debates as monologues by the cultural elites over how they defined noise and whom they considered as noisy. Typically, loudness was associated with the "backward" lower classes and demure quietness was associated with the modern and educated middle and upper classes. From the pro-British *al-Muqatam* to the nationalist Watani and later on Wafdist newspapers, there was a similar "civilizational" message about the urgency of elevating the "tastes" of everyday Egyptians. Silencing or muffling popular voices emanating from the streets was one of the persistent themes repeating not only in the press but also in fiction and nonfiction.[24]

Throughout this book, we will encounter a host of tensions between the cultural and modernizing elites and middle classes and the Egyptian masses over competing and class-defining definitions of noise. The "appropriateness" and the value of silence and public quietness—whether in the streets or in more formal public venues—was also open for "debate." What is good sounding music, and what is bad and cacophonous music? Who decides? Are there occasions, like weddings and funerals, when loudness should be temporarily allowed? How do the sounds of traffic, weddings, and funerary processions change over time? What are the class implications of these changes? What about "sanctioned" sounds emanating from the state or the established religious authorities? How are these official sounds justified? These are just a few of the questions I hope to answer. But before we can begin our discussion, we must briefly address some of our visual biases and a host of common misconceptions about the nature of sound and historical study.

From Ocularcentrism to the All-Seeing State

> The modern sensorium remains more intricate and uneven, its perceptual disciplines and experiential modes more diffuse and heterogeneous, than the discourses of Western visuality and ocularcentrism allow. This is true of the religious dimensions of the sensorium, but it is also true of its Enlightenment valences, even that notorious source of scopic dominion, seventeenth- and eighteenth-century natural philosophy.
>
> Leigh Eric Schmidt, *Hearing Things*[25]

Ocularcentrism reifies sight and visualism as being at the apex of a sensory hierarchy, explaining what is deemed to be the detached, objective "superiority" of Western modernity, while often contrasting it to more "primitive" oral and illiterate cultures. To generalize about how different cultures or civilizations are more intuitive, or have better ears, or are more visual and objective than others is misguided at best.[26] It is overly simplistic to paint with a wide brush some cultures as more aural or visual than others, a view that is not only problematic but more often than not mistaken. Leigh Eric Schmidt's *Hearing Things*, for example, expertly discusses the importance of aurality and listening in evangelical eighteenth- and nineteenth-century America, demonstrating in part the futility of ranking societies along some arbitrary visual-auditory scale.[27] To be sure, there are no objective measures for such broad generalizations, and as is often the case, arguments for cultural exceptionalism almost always end up essentializing and obfuscating more than explaining and revealing any worthwhile "truths."

As recent studies of orality and aurality in Western contexts have revealed, listening and orality are just as important as visual perception in European and American contexts, and the supposed shift to a visual-centric modernity is exaggerated. In the last decade alone, several scholars from various disciplines have credibly contested such dualistic ocularcentric views.[28] When humanists and social scientists have closely examined diverse early modern and modern societies, they discover that "hearing and sound remained critical to the elaboration of modernity." As Mark Smith explains in his most recent work on sensory history, "virtually all the evidence produced by the historians of aurality and hearing of the modern era points to the continued importance of hearing and, implicitly at least, heavily discounts the effect print had on diluting aurality in favor of sight."[29] This rings especially true when we consider the advent of sound technologies, from phonographs and telephones at the turn of the twentieth century to MP3 players and iPhones at the beginning of the twenty-first century. The whole notion of classifying the senses along some objective continuum is futile. There are no objective measures, in the toolkits of humanists and social scientists at least, that can rank which one of our senses is more factual, accurate, or objective. It is striking, though, that visuality became so ingrained as the objective, scientific sense despite quite a bit of evidence to the contrary. The ease with which a practiced magician can trick the eyes of most observers, or the many studies by criminal justice scholars and legal psychologists revealing the unreliability of eyewitness testimony is reason enough for a healthy dose of skepticism concerning the supposed objectivity of the eye.[30]

Although philosophical ocularcentrism died out in the last century, its traces are still strongly felt in the vocabulary we use. While writing this book, I felt somewhat hampered by the relative lack of sound-related vocabulary for analytically articulating the perception and expression of sounded events. At the same time, while sound-related vocabulary is scarce, the usual scholarly vocabulary and also everyday contemporary English is replete with visual descriptors of critical analysis. Scholars examine, observe, spot, survey, scan, inspect, scrutinize, dissect, probe, reflect, focus, visualize, look at, highlight, spotlight, bring to light, underline, accentuate, point out, picture, illustrate, draw out, draw attention to, and so on and so forth. This may explain, in part, the dearth of historical studies examining the soundscapes of the past.

From Foucault's Panopticon to de Certeau's City Streets

Michel Foucault's discussion of the panoptic power of the modern state can be very helpful in understanding the dramatic changes taking place in "modernizing" societies in the nineteenth and twentieth centuries. For the first time, ordinary people en masse were systematically counted, regulated, conscripted, and educated by the state. Within modern Egyptian historiography, Timothy Mitchell's seminal *Colonising Egypt*, fruitfully draws on Foucault's adaptation of the panopticon concept to examine some of the late-nineteenth-century plans drawn up by Egyptian and, later on, colonial authorities. Despite Foucault's philosophical anti-ocularcentrism, his overreliance on ocular metaphors and his articulation of the panopticon as the ultimate conceptual instrument of state power, unintentionally perhaps, makes vision and surveillance into an almost inescapable tool of control and state oppression.[31] To Foucault, seeing is controlling, or as he puts it, "the gaze that sees is the gaze that dominates."[32] Theoretically, it would seem, there is no escaping the all-seeing, godlike panopticon.

Foucault adapted his conceptualization of an all-seeing figurative panopticon from Jeremy Bentham's late-eighteenth-century panopticon prison plans, which featured a circular or semicircular prison building designed to allow visual access to all the prison cells from a centrally located guardhouse. Although it is impossible for the prison guards to watch thousands of prisoners simultaneously, the inmates are not able to tell if and when they are being watched. In theory, this psychologically compels prisoners to regulate their own behavior, since they assume they are always being watched. The panopticon, as a concept, becomes the perfect metaphor for explaining the figurative all-seeing eye of the modern state, with all the measuring, counting, and institutional tools of modernity at its disposal. Curiously, despite the fact that Bentham's original panopticon prison plans provided for listening tubes to "listen" to the inmates, only the visual surveillance was used by Foucault as a conceptual illustration of the ordering powers of the state.[33] On another, more practical though equally somber level, the fixation on panopticons and visual surveillance misses out on the fact that modern states and their enforcement authorities listen as much as, if not more than, they observe and watch over their "subjects." "Surveillance" by the authorities was more concerned with what people said: the jokes they made, the chants they shouted in the streets, the rumors and gossip that were current in cafés and other social circles. Initially, spies and informants wrote down what they heard, and then as new technologies were introduced, conversations and phone calls were intercepted and recorded.

In its broadest application, the concept of a visual panopticon extends to the exhibitions, plans, maps, and surveys of the state. These written and drawn-out surveys, censuses, and maps supply two-dimensional bird's-eye-view plans that are intended to be, though not always are, put into practice. In fact, the blueprints of the state—be they for the building of village schools or for restructuring entire cities—may be perfect in how they are drawn out and in the way they purport to survey, control, and observe, but the reality on the ground is loud, messy, and often contradictory.[34] This is one of the reasons why Henri Lefebvre and Michel de Certeau were critical of Foucault's panopticon schema. The other reason is the seeming inescapability of its visual totalitarianism. Lefebvre felt that in this case, Foucault never bridged "the gap between the theoretical (epistemological) realm and the practical one."[35] In other words, by fixating almost entirely on such visual and printed plans, we neglect their applicability, practicality, and execution. This, according to Lefebvre, "amounts to the overestimation of texts, written matter, and writing systems, along with the readable and the visible, to the point of assigning to these a monopoly on intelligibility."[36] Doing so gives us an incomplete picture at best, and leaves us with very little understanding of what is in fact happening in the streets, or in the "space of people who deal with material things."[37]

De Certeau addresses this disparity between theory and practice by delineating and differentiating more clearly between the definitions of geographic *place* and *space*. Place is the theoretical and geometrically drawn out plan, and "space is a practiced place." I would add, that as a practiced place, space must be embodied and multisensory, as it is by definition inhabited, lived in, used, and walked through. In other words, a *place* is "transformed into a *space* by walkers."[38] Just as importantly, one can also find spaces that are "outside the reach of panoptic power," by figuratively "walking in the city" and familiarizing oneself with its streets and street life on a micro level.[39] In other words, de Certeau's solution is to temper the theoretical gaze of the panoptic "eye in the sky" with a closer more embodied examination of the actual streets, as they are being physically used and walked through by everyday people.

De Certeau's insistence on the importance of walking the city is in effect urging a multisensory understanding of space. By literally or figuratively walking in the streets, one can move beyond the watchful gaze of the state, or beyond the two-dimensional plans and maps that were drawn out and put into place by capitalists or state bureaucrats. By moving closer to the streets, we are able to see, feel, hear, touch, and smell. The lived and embodied spaces and movements of the streets can be discovered, revealing a great deal more about everyday life. In many ways, this book takes a page from de Certeau, as it "walks" in Egypt's urban streets to watch, listen, and sense how ordinary Egyptians dealt with everyday street life. In short, *Street Sounds* is less concerned with the mental and theoretical realm and more concerned with the lived, social, and material realm. It is in this "space of social practice, the space occupied by sensory phenomena" that sounds and soundscapes fill a vital role in defining and shaping public space and street life for everyday men and women.[40] Thus, in critical ways, listening in to the sources, and examining what is happening down in the streets on an intimate sensory level, can counterbalance the distant, almost entirely visual plans and depictions drawn up by state and colonial planners.

Intersensorality: Touching, Smelling, and Tasting History

> The senses are multiply related; we rarely if ever apprehend the world through one sense alone. Indeed, under conditions in which any one sense predominates, closer inspection may disclose that the predominating sense is in fact being shadowed and interpreted by other, apparently dormant senses.
>
> Steven Connor, "Edison's Teeth: Touching Hearing"[41]

Although my primary concern in this book is with sounds and soundscapes, all of our senses are interconnected, and most humans perceive their environment using all five senses, often simultaneously. We cannot talk of sounds and listening without taking into account all of the other senses. Or as Yi-Fu Tuan eloquently puts it: "Most people function with the five senses, and these constantly reinforce each other to provide the intricately ordered and emotion-charged world in which we live. Taste, for example, invariably involves touch and smell: the tongue rolls around the hard candy, exploring its shape as the olfactory sense registers the caramel flavor. If we can hear and smell something we can often also see it."[42] To put it another way, "the sense we make of any one sense is always mixed with and mediated by that of the others."[43]

Social spaces are multisensory because they are both experienced and produced by those who inhabit and use those spaces, using all five senses. To inhabit a space is to physically live in it and embody it by corporally seeing, touching, tasting, smelling, and hearing the movements, textures, flavors, odors, events, sounds, and conversations within that sensory environment. But that, of course, also means that each individual inhabiting that space will also naturally produce and project sensory outputs for others to perceive. By merely existing and occupying any social space, a person is exuding sound and body odor (pleasant or unpleasant), and is visually a part of the tapestry of the environment. In a crowded space, or when greeting familiar faces, touch also comes into play. Taste is shared through some of the common foods and drinks communally consumed in the homes, streets, markets, and cafés. To put it plainly, social spaces do not exist in a vacuum but are simultaneously populated by dozens of sights, sounds, smells, tastes, and textures. Our senses are not only the primary means with which we imperfectly comprehend and digest our immediate environment but also our only means of communicating and conveying ourselves to others.[44]

When we temper our exaggerated reliance on the eye and begin to "listen" more to the historical sources, we begin to perceive things differently. For example, the imagined all-powerful and all-seeing state, with all its new tools of control, is, more realistically, brought down to the level of the street where both its authority and its significant limitations can be more accurately gauged. Also, when we rely exclusively on maps, plans, aerial photographs, and surveys, we stay aloft with a bird's-eye view and neglect the actual streets and the everyday lived realities experienced by ordinary people. When we get a little closer through a multisensory reading of the available texts, we cannot help but listen, and maybe even perceive with our other senses and not just with our eyes. In short, the closer we get to the body and all the senses, the closer we can also get to the experiences and practices of everyday street life, and through them, we can also better comprehend how ordinary people accommodated, appropriated, or resisted state plans and authority.

Ways of Hearing: The Archives as an Earwitness to History

> One problem that came up immediately when I set out to write sonic history was the belief that, unlike a document, sound is ephemeral, going out of existence even as it happens. . . . This comparison is misleading, if not mistaken. Historians do not usually write the history of documents; they interpret the past, all of which has gone out of existence as soon as it came into being, just like its sounds. And like any other experiences, sound and hearing can be partially recovered and interpreted from documents and material culture.
>
> Richard Cullen Rath, "Hearing American History"[45]

For contemporary events and for many historical events that occurred in the second half of the twentieth century, sound and audiovisual recordings are available for inspection. But how can researchers fill in the sensory gap when writing about historical periods before the invention of recording technology? The obvious solution to this problem can be found in the same texts that historians have been using all along. That is, as R. Murray Schafer aptly puts it, historians "will have to turn to earwitness accounts."[46] Even though we visually (and silently) read the texts found in libraries and archives, that does not mean the writers were exclusively depicting visual "observations" in their writings. People absorb and process information, be it physical or cultural, from their immediate environment using all five senses. Lest we forget, the physical act of writing is as much tactile as it is visual, and the information conveyed will inevitably be multisensory.

Historians often have no problem accepting the visual observations made in archival texts and other written sources, yet written reports of sounds and noise have been either neglected or assumed to be less accurate than reports of visual observations. I would argue that this is not a conscious neglect; we are conditioned to elevate the visual above the other senses. Perhaps because as historians we are by definition highly literate, we intuitively privilege the visual and silent understanding of reading and knowledge acquisition. Part of the problem, as referenced in the quotation from Richard Cullen Rath at the beginning of this section, is the mistaken belief that unlike visual observation, aural observation is especially ephemeral—that what we hear dissipates into thin air as soon as it is sounded. But aren't all of our sensory perceptions fleeting, including the supposedly objective and scientific sense of vision? What we "see" at any particular moment is merely the light reflecting upon the object of our gaze and *almost* instantaneously being *read* by our brain immediately after the event has passed. Although light travels quite a bit faster than sound, unless the event we are perceiving is miles away, our mind perceives both the sonic and the visual "echoes" of the event as happening almost instantaneously. So, in fact, both what we see and what we hear are ephemeral moments taking place outside our bodies and later "heard" and "seen" as the light waves and sound waves reach our eyes and ears and are "read," or perceived, by our brains. The same, of course, goes for the rest of our senses, because what we touch, taste, and smell has to be cerebrally processed as well. In this sense, if we are near an event, all that we perceive are echoes of the neurons firing in our brains less than a millisecond after the physical event has taken place.

In other words, writers can document only the memories of what they may have heard, seen, smelled, touched, or tasted. All these sensory impressions, including the visual, are in this sense ephemeral and are recalled and written down after the events have already passed. This is true even if our historical chroniclers always have pen in hand, ready to "instantaneously" document an event. All sensory recollections are processed cerebrally before being recalled and selectively reconstructed via ink on the page. It is often forgotten that written visual impressions are physically and tactilely reproduced by someone's hand as it manipulates the writing implement on the page, and the scribes and chroniclers writing out these records were surely documenting events that they or others experienced multisensorially. If we attune ourselves to the sounds within the historical or literary texts we are examining, we can inevitably see, or rather "hear," what the original authors are conveying regarding the sounds, noises, and speech that they may have heard.[47] Writers and record keepers will inevitably have an abundance of references to what they have heard and will consciously or subconsciously relay that in their prose. Naturally, writers do not record just what they have seen and heard. Depending on the events they are covering and the context of the document they are writing, they can also detail olfactory, tactile, and gustatory information.[48] The senses are all relationally embodied and traces of touch, smell, and taste are usually present in written sources that relate what was seen and heard. When it is relevant and whenever the sources allow, I will relate these senses as well.

Although this next point may seem counterintuitive at first, photographs and other images also do not relay only visual information. They can be full of other sensory information as well and can convey critical information about past sounds and soundscapes.[49] Almost in the same way that texts are full of auditory data so are images and photographs, and if "read" critically and along with supporting documentation, they can be as useful in reimagining some of the lost soundscapes of the past. The Egyptian illustrated press, for example, is full of photographs representing varying aspects of Egyptian daily life. Many of these images have obvious clues to everyday sounds and soundscapes, including depictions of sounded technologies, such as radios and loudspeakers. Photographs of city traffic made up of motorcycles, automobiles, and tramways allow us to easily "imagine" the cacophony that must have been present. Bands playing their drums and brass instruments, children singing and playing, amusement rides set up to celebrate various religious and secular holidays—when used along with the textual records, all these offer the careful and critical historian additional documentation of the sounds that were present at the time. This may seem obvious to us now, yet historians have rarely used images in this way. In the same vein, early silent documentary and newsreel films depicting the Egyptian streets were useful to this study; especially those that were not staged but were set up to "clandestinely" capture Egyptian daily life. When viewed critically and correlated with other sources, a picture may reveal a host of sounds.

Almost all historical documents contain some relevant sensory information, from government and colonial records and police and intelligence reports to photographs, films, sound recordings, newspapers, memoirs, and travel literature. Travel memoirs and writings are bound to contain a rich amount of multisensory references. Unfamiliar sounds, sights, tastes, sensations, and smells are certain to leave an impression on the traveler, pilgrim, or tourist. Of course, historians should be cautious with the inevitable generalizations and stereotypes bound to be recorded when travelers are writing about unfamiliar cultures. Cases from sharia and civil courts and police records should also contain plenty of information about sound, noise, and aurality, from rumors and hearsay heard by spies and police informants to noise complaints filed by angry neighbors. Reading these available sources with our senses attuned to the wealth of sensory information they contain can only enhance and add texture and breathe life into our historical narratives.

Research for this book was mainly conducted in archives and libraries in Cairo and London. I used a variety of archival sources, periodicals, memoirs, diaries, travel books, photographs, films, music, songs, and audio recordings. The Egyptian National Archives (Dar al-Watha'iq al-Qawmiyya [DWQ]) in Cairo and the National Archives (TNA) of the United Kingdom in London were essential, providing a plethora of information on the changing Egyptian infrastructure while also providing intelligence reports on the Egyptian streets. Both archives had a great deal to say about Egypt's changing sounds and soundscapes, though as I will discuss, most of this information was indirect and had to be inferred from the available records. The Egyptian National Archives in particular were vital in giving me a sense of how the Egyptian state tried to silence or regulate unofficial noise, from the setting up of citywide restrictions on public celebrations to several (failed) attempts at regulating the calls of street hawkers. Petitions objecting to the enforcement of these laws and regulations were important in giving a voice to the people, and sometimes were quite effective in tempering the power of the state.

I used the Egyptian National Library (Dar al-Kutub al-Misriyya) mainly for Arabic periodicals. Newspapers and other periodicals are of course loaded with historically relevant, yet still largely untapped, aural and sensory data. Of the many hundreds of Egyptian periodicals that I have read, I have yet to find one issue that does not have a reference to sounds, soundscapes, music, noise, speeches, rumors, gossip, and/or conversations. Diaries, memoirs, and letters were especially worthwhile in my research, as memories are almost always triggered through sensory markers associated with all five senses. When it comes to sensory historical research, travel accounts are arguably the most promising yet also the most problematic of the sources I used. Travel accounts are loaded with sensory information, as foreigners who are unfamiliar with the new sounds, sights, smells, tastes, and textures they encounter rush to write down every unfamiliar detail. They are much more likely than other writers to take note of subtle everyday sounds, ordinarily missed or ignored by local residents because they are so familiar within the immediate audible environment. Foreign travel accounts, however, are also often full of inaccuracies, stereotypes, and varying degrees of ignorant orientalist prose. The quality of this genre of literature varies widely, of course, with some writers much more accurate and well informed than others. Despite their flaws and shortcomings, when read carefully and against the grain, these sources can be invaluable to a historian of the senses.

The Organization of This Book

Temporally, *Street Sounds: Listening to Everyday Life in Modern Egypt*, covers the period from the later years of the nineteenth century until the middle of the twentieth century. Conceptually, the book is divided into three parts. Part I examines the mundane sounds of Egyptian street life and the inevitable contestations between everyday people and the state over "ownership" of the streets. Part II focuses on the sonic implications of modernizing infrastructure, electrification, and increasing traffic on Egyptian urban life. Part III examines the sounds and impact of public spectacles, parades, and processions.

The chapters within these parts are thematic in focus, highlighting major technological and cultural transitions taking place as they relate to the changing soundscape of the Egyptian streets. Inspired in part by Michel de Certeau's *The Practice of Everyday Life*, in Chapter 1, I describe life in early- to mid-twentieth-century Cairo from a pedestrian's perspective. "Walking the City: Street Voices, Traffic, and the Mundane Sounds of Everyday Life," is devoted to the sounds of pedestrians and commuters as I analyze the ways that they used, occupied, and walked through public spaces to commute, work, sell, and shop, and to entertain and be entertained. The first part of the chapter examines some of the social implications of embodied sounds, from jingling anklets and bracelets to footsteps, as ordinary Cairenes were negotiating their way through a rapidly changing city. The second half of Chapter 1 focuses on the calls of street hawkers, entertainers, and merchants who relied on their voices to advertise their goods and services. Chapter 2, "Silencing the Streets: Classism, Fear of the Crowd, and Regulating Sounds and Bodies," discusses both the Egyptian government's attempts at regulating and silencing public spaces and the class implications of these policies. New anti-begging and anti-homelessness discourses invoked fears of an imminent breakdown of public order and even public health. Here I also document the interrelated, and ever-present class bias in the Egyptian press, especially with regard to its coverage of the urban streets, street hawkers, and the itinerant poor.

Chapter 3, "Roads and Tracks: Modern Traffic and the Sensory and Social Impact of Trams and Automobiles," considers the modernization of Egypt's urban infrastructure, especially roads and tramway tracks. The changing sounds and the social impact of growing urban traffic are carefully examined, with an emphasis on the introduction of tramways and motor vehicles. The rest of the chapter documents and elaborates on the sonic impact of new urban spaces, from large city squares to bustling transportation hubs. In particular, the problems of dramatically increased motor traffic and early attempts at regulating car horn noise are investigated. Chapter 4, "The Soundscapes of Modernity: Electricity, Lights, and the Sounds of Nightlife," begins by examining the professional lighting of Egyptian cities by private gas utilities and the introduction of electricity and electric lighting. The sonic implication of electrification was of course enormous, as it not only allowed the eventual proliferation of radios, loudspeakers, and tramways, but just as importantly, electrification also led to the introduction of electric lights, which forever changed the sounds of the Egyptian night. The growth of a regular and boisterous nightlife, and the establishment of newer places of public leisure, from amusement parks to cabarets and movie theaters catering to diverse audiences, will be closely examined.

The last two chapters, deal with the sounds of public spectacles, which contrary to their label, were as auditory as they were visual. Chapter 5, "The Sounds of Weddings and Funerals: From Brass Bands to Wails and Ululations," examines the evolving street sounds of traditional Egyptian weddings and funerals, which involved multiple and elaborate street processions and a variety of audible and visible displays. It considers the changing roles of street music, singing, loud funerary grieving, and other unique and important verbal and nonverbal vocalizations. I conclude the chapter by examining some of the ideological attacks against embodied and auditory aspects of these traditional ceremonies by the government and by both secular and Islamic modernists. The one point that all the "modernizing" camps agreed on was their belief in the general ignorance of the vast majority of the population and the urgent need for education, reform, and uplift. Class distinction was especially important for Egypt's growing middle classes, as they self-consciously attempted to define and separate themselves from the masses. Discourses of sensory differentiation are almost always invoked for the purposes of class distinction.

In Chapter 6, "Sounding Out State Power: Cannons, Music, and Loudspeakers," I delve into the Egyptian state's appropriation of large religious and secular celebrations and festivals. This chapter centers exclusively on the official sounds and spectacles performed and sponsored by the Egyptian state in an ongoing effort to legitimize its secular and religious authority in the eyes and ears of the masses. I demonstrate that drums, cannons, and twenty-one-gun salutes are as important as uniforms, flags, and propaganda posters. I will particularly focus on the state's use of music, microphones, and radio speeches broadcast over loudspeakers in parades, festivals, and other large public gatherings.

Although *Street Sounds* is primarily a history of the changing sounds and soundscapes of late-nineteenth to mid-twentieth-century Egypt, it also reveals a great deal about ordinary people and their interaction with their social, economic, and cultural environment. By "listening" to the changing street sounds during this period, I intend to document the "voices" of everyday people who made themselves heard through a variety of inventive ways. Everyday Egyptians ululated and beat on drums to celebrate, and shrieked loudly to mourn their dead. They spoke out, sang, and chanted, and in the process they sounded out their thoughts, changing values, habits, devotions, and emotions to one another and in reaction to the forces of modernity and state authority. The streets themselves became de facto battlefields or sites of negotiations between the state's attempts at ordering and "modernizing" society, the middle and upper classes' judgmental discourses about top-down, ordered civility, and the urban masses' accommodation of and resistance to both. *Street Sounds* is about the intersection of street life with modernity, class formation, and the persistent attempts by the Egyptian state to control the urban streets. As I intend to demonstrate in the following pages, an embodied history in tune with the multisensory and varied ways that everyday people engaged with their immediate environment can provide us with an intimate street-level analysis of the shifting realities of everyday life.

Mundane Street Sounds and Ownership of Public Space

Egyptian everyday life in the many quarters of Cairo is rich with sonorous artistic expressions. A simple and innocent art perhaps, but it is still art nonetheless. For all those who are in the streets are dancing or singing or dancing and singing together. The sellers of grapes, guava, and apricots for example are singing a sweet song describing their goods. The Ramadan town crier [*misaharati*] is singing, the call to prayer [*adhan*] is a song, and Qur'anic recitation at funerals, weddings and religious festivals sound like melodic hymns. Street religious festivities and parades play music and sing aloud . . . and so do the wedding and circumcision processions that pass on a weekly basis in our street and in neighboring streets.

Fathi Radwan, *Khati al-Ataba*[1]

While not wishing to dwell too much on the volume of work of the *qisms* [police stations], I will quote the following figures from one *qism*, 'Abdin, to show what a maximum effort is required to obtain a minimum result in checking the daily nuisances of the town. During the year 535 contraventions were made for vagabondage, 448 contraventions were made for beggars, 2,398 contraventions were made for hawkers, and 731 contraventions were made for illegal tram-riding.

Ministry of the Interior Report for the Year 1922[2]

1

Walking the City
Street Voices, Traffic, and the Mundane Sounds of Everyday Life

> The ordinary practitioners of the city live "down below," below the thresholds at which visibility begins. They walk—an elementary form of this experience of the city; they are walkers, *Wandersmanner*, whose bodies follow the thicks and thins of an urban "text" they write without being able to read it. These practitioners make use of spaces that cannot be seen; their knowledge of them is as blind as that of lovers in each other's arms.
>
> Michel de Certeau, *The Practice of Everyday Life*[1]

DE CERTEAU'S DEPICTION of street life "down below," captures an embodied, extra-visual sensory knowledge that ordinary urban dwellers have of their immediate environment. To a city resident, there is an innate sensory familiarity in a neighborhood street, supplementing known sights with equally familiar pleasant or unpleasant sounds, smells, and tastes. Even the touch and feel of specific dips, curves, and cracks in the streets and sidewalks become intimately and blindly familiar to pedestrians as they daily take the same route to work, school, or the market. As I demonstrate in this chapter, by "listening in" to street life, we can partly reveal how people lived their everyday lives and, more importantly, clarify how they dealt with state authorities in their mundane struggles over the use and ownership of the public streets. Compared to other approaches, a sensory approach is also more embodied and intimate, as it sheds more light on how ordinary people adapted to the plethora of modernizing changes taking place during the first half of the twentieth century. The Egyptian streets were not only a living, breathing laboratory for rapidly unfolding infrastructural and technological transformations but also one of the few places where most elements of society could, potentially at least, interact face to face.

In this chapter, we will be listening in to these embodied, often mundane exchanges, while documenting the dramatic changes in street sounds and soundscapes, which accelerated as the twentieth century roared on. The unprecedented infrastructural and technological changes taking shape at the time touched every aspect of Egyptian daily life, irreversibly changing how people's immediate environment was perceived by their senses. In this chapter, I expand on this analysis by examining how these spaces and new transportation technologies were used, inhabited, and embodied by everyday people to commute, work, sell, and shop and to entertain and be entertained. In other words, this chapter is about the men and women that used, occupied, and walked through the streets on a daily basis, from carriage and tram drivers to pedestrians, street hawkers, merchants, and beggars. When the sources allow, I focus on the body itself and the sounds that people made as they walked, moved, and even gestured. Most of these bodily sounds were not vocalized or played through some sort of instrument, but the very act of walking when aided with footwear or, for women, wearable accessories like anklets and bracelets can, intentionally or unintentionally, project meaningful and often evocative sounds. In the second half of the chapter, I focus on Egyptians who made a living in the streets—from water sellers and beggars to merchants and street hawkers—all of whom left their sonic imprints on the streets. Street hawkers in particular, who are still a prominent part of Egypt's soundscape, feature in this chapter and the next—as do their frequent battles with the state.

Pedestrian and Traffic Noise in Old Cairo

In the Introduction to this book, I discussed a few ways that historians can employ to examine past sounds and soundscapes through the careful reading of texts and even photographs. Indeed, the bulk of my sources consists of texts and will require "ear-witnessing," as described by R. Murray Schafer.[2] However, good fortune sometimes intervenes, as when I discovered some very useful early recorded sounds. The most interesting example is the raw footage of an early sounded newsreel, or documentary film, recorded in Cairo by the Fox Movietone company. Shot on December 28, 1928, the unnarrated, fifteen-minute film segment depicts a variety of street sights and sounds in two primary locations: the camel market in Imbaba, northwest of Cairo, and al-Muʿiz Street near the Husayn district in old Cairo (Misr al-'Adima), part of the original core of historical Cairo. As I am primarily concerned with the sounds of pedestrians and urban traffic, I will mainly focus on the soundscapes of al-Husayn district.[3]

In all the recorded scenes, a large microphone was placed a few feet away from the camera, on a four-foot stand by the main street. The camera captured the microphone on a few occasions, as it panned out to film passersby (see Figure 1.1). Some of the pedestrians, street merchants, and carriage drivers, are too busy to notice the camera or microphone and simply walk by. A brass shop owner and his customers, for instance, do not show concern and appear largely oblivious to the camera filming them across the street. The microphone easily picks up the sounds of the brass containers and plates as they click against each other when customers handle them. Murmurs of conversation between the buyers and sellers can also be heard when no automobiles are passing by, drowning all other sounds with their loud motors and incessant honking. Mostly, however, passersby make sure to be filmed as they "casually" slow down or momentarily stop and glance up to look at the camera and microphone. Later in the film, the film crew allows some of the crowd to gather around the camera, with the most daring in the crowd performing for the camera. The tourist guide (dragoman), who was hired by the film crew, shows particular eagerness to dominate these scenes, sometimes shooing away men and women who block the camera's view of him. At other times, he participates along with others in performing for the camera by talking to and directing those gathered around him. Comfortable in the knowledge that the American film crew does not speak a word of Egyptian Arabic, he even yells out a dirty joke and later ritualistically trades insults (in the 'afiya tradition) with several men off-camera.[4]

What intrigues as much as the conversations, jokes, and insults are the other, more mundane background sounds being captured through the microphone. There are plenty of clearly audible traffic noises from bicycles, automobiles, a galloping horse, and donkey carriages and carts, along with a few horse-powered omnibuses, as they attempt to thread their way through the traffic and the crowds of pedestrians. All of these sounds complement the steady sounds of conversations, whispers, and murmurs throughout the film. As can be also seen, although most of the main thoroughfares of al-Husayn were paved by 1928, the streets are quite narrow, with limited or no sidewalks. As expected, al-Mu'iz Street with all its shops and the Husayn district, right outside the Khan al-Khalili bazaar and al-Azhar University, are bursting with pedestrian and vehicular traffic sharing the same thoroughfare. These communal and competing uses of the streets necessitate yet another layer of various rings, honks, shouts, and even knocks and thumps in order to warn pedestrians to yield or part way for the incoming vehicles.

FIGURE 1.1. Street Microphone.
SOURCE: *Cairo street scenes—outtakes*, Fox Movietone News Story, December 28, 1928 (University of South Carolina, Moving Image Research Collections). Copyright holder: University of South Carolina.

Although in 1928 automobiles were still not as numerous as animal-powered carts and carriages (especially in the Husayn quarter), when one occasionally passes by in the film, the distinctive roar of its engine drowns some of the other street noises. As we would expect, these noises supplement a steady and repeating blaring of the horn. For example, before a car can be seen passing in front of the camera, its horn is heard for a full twenty-five seconds, repeating rhythmically the entire time, and the driver continues to honk as it passes the camera and the microphone, driving away into the distance. In fact, as can be clearly seen in the film, the right hand of the driver never lets go of the horn, which is attached on the outside of the car's door (see Figure 1.2).[5] It is important to be aware that had this camera crew shot and recorded on the very same day scenes of the wealthier, less densely populated (at the time) Azbakiyya and Ismailia districts, with their large sidewalks for pedestrians and much wider boulevards and squares, the sounds of the streets would have been quite different. In these newer parts of Cairo, depending on the time of the day or night, the sounds of automobile engines and electric trams would probably have dominated.

FIGURE 1.2. Hand on Car Horn.
SOURCE: *Cairo street scenes—outtakes*, Fox Movietone News Story, December 28, 1928 (University of South Carolina, Moving Image Research Collections). Copyright holder: University of South Carolina.

One of the unexpected sounds for me was the prevalence of bicycles bells, constantly ringing in both al-Husayn and the adjacent al-Muʻiz Street. Some bicycle bells are heard in the background, while a few others are heard and seen as the bicyclists thread their way through the crowds and traffic. On paper at least, as we will see in the next chapter, ensuring that pedestrians were warned of oncoming bicycles was important enough to the authorities in Cairo that as early as 1894, they required all registered bikes to have "a bell or a horn to alert pedestrians."[6] The noise laws regarding animal-drawn vehicles were less straightforward and somewhat contradictory. Whereas horse-drawn omnibus drivers had to be equipped with a "whistle to alert pedestrians and other vehicles,"[7] horse and donkey cart drivers "must drive their carriages as quietly as possible and have to keep their vehicles on the right side of the road."[8] Returning to the 1928 film footage, there is no confusion about loudness and noise when it comes to a cart driver and his assistant who can be loudly heard a few seconds before appearing on camera, carrying a large kerosene fuel tank on their mule-drawn cart. To warn the pedestrians, as they are expertly darting

FIGURE 1.3. Cart Driver and His Assistant.
SOURCE: *Cairo street scenes—outtakes*, Fox Movietone News Story, December 28, 1928 (University of South Carolina, Moving Image Research Collections). Copyright holder: University of South Carolina.

and zigzagging their way through the crowd, the driver bangs loudly on the tank behind him with a stick and his assistant, sitting to his left, is shouting out to the pedestrians in front of him: "Watch out, mister [*effendi*], get out of the way you two!" (see Figure 1.3). When he passes right in front of the camera, he notices the film crew, looks straight into the camera, smiles, and yells out quickly to the tour guide, in jest, "Take us with you, 'O Abu Ali.' Take us so we can work for you!"[9]

Even though this movie somewhat captures a slice of a 1928 winter day in old Cairo, it is vital to remember that in addition to the specifics of a location, *time* is obviously a critical component in how a space is sensorially experienced.[10] A useful historicizing exercise is to imagine how the sights and sounds at these very same locations were different in 1880 and in 1950. Some of the factors that come into play in this calculus include population size and density and whether blaring radios and noisy trams, buses, automobiles, and motorcycles are absent or present and in what quantities. Examining a similar scene occurring in the same location a half-century earlier may provide a fruitful comparison.

When a French traveler named Gabriel Charmes visited Egypt for approximately five months in the early 1880s, there were no trams, automobiles, or bicycles. However, there were a few remarkable similarities between some of what he experienced and the 1928 film. Despite the fact that horse-drawn coaches and carriages were not a common sight in Egypt until the mid-nineteenth century, Charmes was quite impressed with the dexterity and driving abilities of Cairene carriage drivers.[11] According to Charmes, Cairene coachmen "never touch in passing, even when they turn at a sharp angle in the streets singularly narrow, even when they pass amidst, without overturning them, a dozen little hawkers with their almond cakes, their dates, and their preserves spread out on the road before them. A Parisian coachman would smash everything." Cairene pedestrians also earned Charmes' respect, as he was impressed by their dexterity in negotiating the streets and especially traffic: "One may constantly see children crossing under the horse's chest, escaping from under the wheels on the point of crushing them, and avoid by a dexterous movement being run over by a rider at full gallop."[12]

An English traveler, who was in Cairo three decades after Charmes, was a bit less generous with his assessment of both Cairene pedestrians and carriage drivers. After describing how the "traffic of carriages, carts, porters, native women, lemonade-sellers, and sheep was inconceivable," he declares that the sidewalks and streets were so crowded that they "would congeal if it were not for the *arabeah* [carriage] drivers, who charge them like snow-ploughs, crying, '*owar* [*iw'a*] *riglak*' (watch out for your leg!)"[13] Hyperbole aside, because of the density of the crowds and the traffic, automobiles, trams, carriages, and even bicycles had to sound out their presence in a variety of ways to warn pedestrians and other traffic. This merely added to the multiplicity of sounds, old and new, that characterized the urban soundscape. These varied sounds, and the way that drivers, street hawkers, and pedestrians physically and sensorially negotiated the streets, by obeying and—more importantly—disobeying traffic laws, present a mundane, yet clearly more embodied and realistic manner in which everyday passersby used the city streets.

Walking the City

> Surveys of routes miss what was: the act itself of passing by. The operation of walking, wandering, of "window shopping." . . . The act of walking is to the urban system what the speech act is to language or to the statements uttered. At the most elementary

> level, it has a triple "enunciative" function: it is a process of *appropriation* of the topographical system on the part of the pedestrian (just as the speaker appropriates and takes on the language); it is a special acting-out of the place (just as the speech act is an acoustic acting-out of language). . . . It thus seems possible to give a preliminary definition of walking as a space of enunciation.
>
> Michel de Certeau, *The Practice of Everyday Life*[14]

As Michel de Certeau eloquently describes in *The Practice of Everyday Life*, city dwellers have an elemental, sensory, and intuitive knowledge of the city streets and neighborhoods they daily walk and inhabit. As alluded to in the epigraph for this section, walking through and using and misusing the city streets is in itself an act of appropriation. People take ownership of the streets when they use them, whether they are simply walking through, sitting in cafés that extend the seating into the sidewalk, or selling their goods and services in the city's squares and roads. This appropriation and embodied possession of these public spaces was not always uncontested. As I examine in more detail in the next chapter, the police and the political authorities always attempt to limit, control, and regulate the use of public roads. Of course, walking in the streets is a necessarily embodied and intimate experience, in which all of the five senses are consciously and, even more, unconsciously and simultaneously engaged. To navigate a busy road in 1930s or 1940s Cairo, for example, pedestrians must watch, listen, and feel, while also processing (if necessary) olfactory data that is simultaneously bombarding their senses. The smell of fresh baked bread, barbequed corn, or roasted peanuts may tempt or sidetrack a walker to momentarily stop, make an impulse buy, and eat while continuing to his or her destination. The smell of garbage, a tannery, or a slaughterhouse would naturally have the opposite effect—prompting a pedestrian to walk more rapidly and later on perhaps to avoid the area altogether.[15] Horse, donkey, and other animal dung is another forgotten, yet ubiquitous, foul-smelling reality that littered most urban areas before the mid-twentieth century, and which made a walk in any busy urban street a virtual obstacle course with lingering sensory consequences.

Crossing the street was a drastically different experience after tramways and automobiles were introduced at the turn of the twentieth century. In most parts of Cairo and Alexandria, trams did not have separate and fenced tracks, but instead shared the urban streets and intersections with pedestrians and all other wheeled and un-wheeled traffic. Automobiles, motorbikes, and bicycles joined the fray in the first decade of the twentieth century, drastically increasing in numbers in the coming decades. For pedestrians, negotiating these various new and old vehicles with varying speeds, sizes, and sounds, required a fluid, sensory awareness of their immediate environment. They needed this awareness especially at night, when street lighting was still evolving from gas to electricity, and the early electric-lighting technology did not produce light as bright and efficient as it would become by the late 1920s. Keeping an eye, "ear," and "nose" out for road obstacles, traffic, and other pedestrians was a necessary skillset for most city dwellers. By the early 1930s, a pedestrian had to instantaneously and simultaneously feel, watch, and listen for oncoming traffic, which at the time consisted of cars, buses, motorcycles, trams, bicycles, and horse-, mule-, donkey-, and even human-drawn carts.

Indeed, the rush-hour transportation soundscape of Cairo and Alexandria during this era was multilayered and complex. A seasoned commuter or pedestrian was well attuned to the varying sounds and their fluctuating loudness, pitch, and frequency depending on the time of day or night. The electric hums of the tramway and its distinctive bells that were different from bicycle bells, the increasingly ubiquitous car and bus horns and engines, galloping horses, and the continuous jingling of the small bells typically attached to donkey and horse harnesses added to the din. In addition, pedestrians did not just experience these multisensory activities; they in turn produced their own sensory outputs by their mere physical presence and activity.

Footsteps and Other Embodied Noises

> Their story begins on ground level, with footsteps. They are myriad, but do not compose a series. They cannot be counted because each unit has a qualitative character: a style of tactile apprehension and kinesthetic appropriation. Their swarming mass is an innumerable collection of singularities. Their intertwined paths give their shape to spaces. They weave places together. In that respect, pedestrian movements form one of these "real systems whose existence in fact makes up the city."
>
> Michel de Certeau, *The Practice of Everyday Life*[16]

Here, Michel de Certeau is primarily concerned with the "swarming mass" of footsteps—steps that create "intertwined paths" that "weave places together," and are a critical component in what makes up the actual city.[17] In this way, pedestrians—and also I would argue, drivers, commuters, and especially street merchants and even beggars—appropriate and, in some ways, physically possess the streets. In the process, they make those streets their own. De Certeau's analysis of the importance of actual footsteps in creating real and embodied systems of circulation does not directly invoke sound. However, since we don't live in a literal vacuum, sound is a necessary by-product of any tactile and kinesthetic action, including walking and footsteps. I would argue, in fact, that all sounds emanating from pedestrians, from conversation or other vocalizations to the actual sounds of footsteps, add yet another layer to people's appropriation of space. As Hillel Schwartz eloquently states, "think of pavement as a rough sound ribbon whose recording and playback mechanisms are shoes, hooves, wheels, canes, pelting rain, hail, and other falling objects."[18] I would also add that in some contexts, even an individual's footsteps could invoke direct meaning to those who are within earshot or visual range of the passerby.

But how much audible sound can one pedestrian make? In a crowded street during the middle of rush hour, not much, as any single person's footstep would undoubtedly be drowned out in the middle of all the noises emanating from the machines and the crowds. On a quiet afternoon, pedestrians' footsteps can be heard, even more so if they want to call attention to themselves by increasing the amount of sound emanating from their walk. Nonvocal auditory cues can easily be projected by walking a little more forcefully or picking up the pace, which could alert another pedestrian walking ahead of them of their impending approach from the rear. This auditory clue allows the pedestrian in the front, without looking or turning around, to not only detect another person's approach but also to anticipate when that person will catch up and want to pass, allowing the person in front to move aside if needed. Making one's presence known while walking would be particularly useful at night, especially in poorly lit areas, since walking too silently may be viewed with suspicion. Making deliberate noise while walking would also decrease the chance of being startled by or startling someone else in the middle of the night. The *ghafar* (sg. *ghafir*), or "night guards," who were much more plentiful in Egypt's urban and rural streets at the turn of the twentieth century, regularly relied on projecting some sound and purposively making their steps heard as they walked their designated beats.[19]

The characteristic sound of a *ghafir*, however, was not his audible footsteps but his regular nighttime cry, which in the stillness of the night could be heard for quite a distance. To the uninitiated ear, the *ghafir*'s calls had a mysterious eeriness to them, as a visitor to Cairo in the late nineteenth century reported: "every quarter of an hour they utter a cry, which everyone ought to repeat after the one that precedes him, and which, in being prolonged the whole length of the street, reaches the next, and is thus spread throughout the district. It is by these means the guardians prove that they are awake. . . . These noises, which succeed regularly till morning, produce in the silence of the night a mysterious impression."[20] The traditional nighttime call of these patrolling beat officers was a short, nonverbal but loud vocalization, followed by "*min hinak*" (who's there!)[21] A *ghafir* used this call to scare off potential burglars, communicate with other guards by creating a sort of sonic security network spreading out to the rest of the neighborhood and beyond, and perhaps to signal wakefulness and readiness to the local *shaykh al-ghafar*. These nighttime cries were pervasive in Egypt well into the twentieth century.

There were other reasons for individuals to consciously or subconsciously project the sounds of their footsteps. The most common reason, perhaps for both men and women, was to call attention to themselves, either to affirm their social or economic standing relative to those in their immediate surrounding or to flirt and grab the attention of a potential mate or love interest. In the first few decades of the twentieth century, a significant proportion of Egypt's population could still not afford a pair of shoes. Many of the Egyptian urban and rural poor and most of the beggars and street children and many of the street hawkers walked barefoot, the clearest physical indication of someone's lower socioeconomic status. So, until shoes and other footwear became cheaper and more available (around the middle of the century), simply wearing a pair of shoes signaled at least a basic level of economic subsistence. Even to this day in Egypt, figuratively describing someone as *hafi* or *hafiyya* (barefoot) is considered an insult, indicating not necessarily poverty but an uncultured vulgarity. For those who could afford footwear, the level and type of sound they could make was somewhat amplified and also dependent on the type of footwear the walker was wearing, adding to the embodied rhythm of pedestrians as they walked about town. The type and price of the shoes a person wore also audibly and visually marked the walker's social status.

The *khuf* slippers worn by traditional Egyptian men and some women at the turn of the twentieth century were made of leather but did not have a heel. The soles were either hardened leather or wood, and the slipper completely covered the foot in the front but was open in the back. Because they were open in the back and not completely fitted around the heels and ankles, they made a distinctive double sound as first the *khuf* hit the pavement and then the bare foot and heel slapped on the leather or wooden material of the *khuf*. The heelless and toeless Egyptian *shibshib* slippers were more practical and mostly worn by Egyptian women. Like the *khuf*, they also traditionally had leather or wooden soles and made a similar double slapping sound as a person walked. The traditional *qubqab*, wooden soled clogs, were more typically worn in the public baths and made a distinctive tap or thud, depending on the makeup of the walking surface. The *khuf*, *qubqab*, and traditional *shibshib* were then completely replaced by the much cheaper plastic and rubber-soled *shibshib zanouba* (or just *zanouba*) starting in the mid-1950s. Because of cheapness of the *zanouba*, wearing them publicly in the city streets was always associated with the working class and the poor.

A well-crafted leather shoe, along with a western suit and a fez, was an essential part of the uniform of the middle-class effendi. For men, a walking cane could have been another visual and auditory marker of economic status, as its hard brass or wooden tip tapped on the pavement or concrete. It was not uncommon for even younger twenty- or thirty-year-old effendis to carry and walk with a finely crafted ebony walking stick. Some of these canes may have been etched with a bit of gold, perhaps in the form of the owner's initials, to signal his wealth or upward mobility. Just like the male effendiyya, by the 1920s, many upper- and middle-class Egyptian women were wearing more western-style clothes and shoes, some with high heels that visually and sonically signaled the wearer's class, education, and "modern" outlook.[22]

Listening to the Jingles: The Sounds of Jewelry and Other Accessories

> The very typical fellahin women of the Delta jingled from ankle to forehead; their foreheads were covered with gold chains and crescents.
>
> Douglas Sladen, *Queer Things About Egypt* (1911)[23]

For most Egyptian women, jewelry was a marker of class, a functional inflation-proof bank account, an economic hedge against divorce, and for some, simply a signal of their gender or sexuality.[24] Perhaps in part because of the varying veiling conventions practiced by many Egyptian urban women at the turn of the twentieth century, middle-class and lower-middle-class women wore, perhaps in addition to high heels, female jewelry and some clothes that appealed to the ear as well as the eye. Most of the jewelry was purposively designed to jingle and tinkle as the wearer walked by. The sounds of high heels or wooden clogs hitting the pavement or sidewalk or the jingling of chains, anklets, or bracelets aurally signaled to all that the passerby was a woman.[25] Just like the sounds produced by an effendi's cane tapping on the pavement, all of these sounds were embodied sounds, since the clothing, shoes, and accessories were an extension of the body and their rhythm and loudness could be consciously or subconsciously adjusted or calibrated by the pace and the movement of the pedestrian. Anklets and bracelets in particular, which were a popular accessory at the time, made a distinctive clinking sound as the wearer walked by. Curiously, women's ankles, anklets, and sometimes most of their lower legs were typically exposed, despite the fact that the wearer usually had her head covered and wore some sort of face veil. Bracelets, sometimes as many as a dozen or more worn on each arm, must have made a different rattle as metal clanked against metal for an additional sonic effect (see Figure 1.4). It was not just jewelry that made an audible fashion statement; some women wore clothes, veils, and traditional wrapping sheets (*milaya laff*) embroidered on the edges with small metal coins, sewn close enough to each other to maximize their jingle.

FIGURE 1.4. Guava Seller.
SOURCE: The author's personal collection of photographs (photograph by Lehnert and Landrock, 1920).

Not surprisingly, many of the slippers, shoes, and accessories discussed here had names that were onomatopoeic, mimicking the sounds that the objects made. This alone is a testament that sound was an integral part of the way jewelry and accessories were used. *Qubqab* (which is pronounced "'ub-'ab" in the Egyptian vernacular) is an obvious onomatopoeia, simulating the sound the wooden clog makes as it taps on the floor. Like the flip-flop, its modern equivalent in American English, the *shibshib* ("shib-shib") similarly sounds out the double noise made when the bare foot slaps the leather sole of the slipper. As for jewelry, both the Egyptian vernacular word for anklet, *khulkhal* ("khul-khal"), and the word for bracelet, *ghuwaysha*, sonically mimic the jingles and metallic sounds created by the walkers' footsteps as the small metal pieces shake or clink against each other. More significantly perhaps, the words *makhtara* (walking gracefully with a swinging gait), *shakhla'a* (walking with revealing, over-accessorized clothing), and *shakhlala* (flirtation) are all onomatopoeia connected to the jingling, rattling, and tinkling of jewelry and accessories.

Men, Women, and the Sounds of the Market

> The kodaker [i.e., photographer] will have plenty of time to photograph the humors of the Musky [Muski Street] even if he is driving, for the traffic is never untangled. . . . There are almost as many stalls as there are shops, though there is no room in the street at all; the most popular are the spectacle stalls—spectacles are almost as much part of the costume as watches and chains in Egypt. Stalls for nougat, Turkish delight, Arab sugar, small cucumbers and oranges, lemonade, boots and shoes, idiotic cutlery, coffee cups and glasses, turquoises and mousetraps are their nearest rivals.
>
> Douglas Sladen, *Oriental Cairo* (1911)[26]

Douglas Sladen, a brash and sensationalistic English academic and travel writer, wrote several books on traveling in Cairo and Egypt around the first decade of the twentieth century. Although his books exemplify orientalist travel literature and also demonstrate Victorian racism, he was a keen observer of people and places, and if read carefully (and against the grain), there is much to be gleaned from some of his narratives and descriptions. In the excerpt that begins this section, Sladen describes Muski Street, which was one of the main shopping areas of Cairo, connecting al-Ataba al-Khadra Square (the city's main tramway hub) with al-Husayn district, where the Khan al-Khalili bazaar, al-Husayn Mosque, and al-Azhar University are located.

One particular market encounter that Sladen had with a veiled merchant woman is worth a closer look. Shopping at Cairo's brass market secondhand stalls, Sladen, along with his Egyptian guide and translator, would often stop and haggle with the female owner of the "chief stall" in the market. Sladen describes the merchant as a contemptuous, "old, awful and closely veiled" woman. After admitting that she was a "great character," Sladen details how she "sat in a recumbent position, stretching out towards the customers her solid but shapely bare ankles in a pair of very heavy gold anklets. They may have been gilt brass, but they were at any rate very handsome. Her bracelets, of which she wore many, were certainly gilt brass." This exemplifies the wearing of anklets and multiple bracelets that was very common at the time, and in the case of the brass merchant, when she moved her arms while walking or gesturing, her multiple bracelets produced a loud jingling sound. Her heavy gold anklets, however, were unlikely to project a sound, unless there were other anklets rubbing and clinking against them; they were meant to visually display her wealth for all to see. Few could afford thick, solid-gold anklets, and most women wore thinner chain anklets, usually with small copper or gold coins and/or crescents that were meant to project a tinkling or jingling sound. Sladen's further description of the "old" lady also reveals a certain grudging admiration for the force of her personality:

> What you noticed most were the extraordinary, expressive eyes between her face-veil and head-veil. She was generally smoking, almost always contemptuous, and flew into fierce passions on small provocations. Sometimes she did not want to have anything to do with a dog of a Christian who dared to dispute the extravagant prices she would put on. She would yell at me and almost throw things at me if I attempted to bargain with her. At other times she would let me put my own prices on anything in the extensive stock she spread on the ground.[27]

It is clear from Sladen's description that he is not quite sure what to think of the brass merchant. When describing her, his usual dismissive, sexist, and racist remarks are mixed with a grudging respect, maybe even admiration with a bit of fear. It is clear, however, that this little episode counters most of the other stereotypes about veiled Egyptian women circulating at the time, and in writing about it, Sladen seems quite uncomfortable or at least taken off guard. Unlike most other travelers who observed veiled women walking in the street from a distance, Sladen actually conversed and dealt with the merchant face to face, listened to her voice, and engaged with her as a person. Had Sladen dug a little deeper, he would have learned that this brass merchant was one among many female Egyptian store or café owners and operators. Known as *muʿallimat* (sg. *muʿallima*) in the Egyptian vernacular, these female merchants and storeowners were a respected part of the urban Egyptian community for hundreds of years. As Evelyn Early, who conducted an ethnography in 1970s Cairo, relates, the "*muʿallimat* are tough women with mannish demeanor who run shops or cafés. Respected and feared by all, they remain untouched by innuendos of loose morals reserved for some women in public places. A *muʿallima* talks sternly, even roughly, with customers and curses as if she were a man."[28] One of the main reasons the *muʿallimat* were relatively free from "innuendos of loose morals" was their relative wealth, as we shall soon examine. This exception was not made for the more economically marginalized female street hawkers and beggars.

Like Sladen, most foreign visitors who traveled to Egypt during the first quarter of the twentieth century were just as obsessed as contemporary observers over what Egyptian women were/are wearing. Typically, foreigners associated the fashion choices of Egyptian women with the marginalization, seclusion, and oppression of women, and the blame was squarely placed on Islam. This was, of course, a gross oversimplification and exaggeration. Gender inequality was certainly a serious issue in Egypt at the turn of the twentieth century, just as it was in the rest of the world, though it is important to historically contextualize this inequality with what was actually taking place in the streets. In reality, like the brass merchant, most rural and urban Egyptian women, regardless of what sort of clothing they wore, participated in a variety of ways in public life, by toiling in the fields, selling and buying goods and services in the markets and streets, and working in homes, factories, hospitals, and schools. This was in addition to running their own households and, for the vast majority who could not afford to hire domestic help, doing most of the household chores. As far as what women actually wore in public, that varied dramatically based on class, region, ethnicity, and occupation. The rapid cultural changes taking place during the first half of the twentieth century, which were visibly reflected in women's fashion, complicated matters. The different manifestations of the face veil worn mainly by urban women would disappear almost entirely by mid-century; head coverings would also mostly disappear in Egyptian cities by the late 1950s, only to make a reappearance in a different more colorful form in the 1980s.[29]

The Water Sellers' Disappearing Cries

> Perhaps no cry is more striking, after all, than the short and simple cry of the watercarrier. 'The gift of God!' he says, as he goes along with his water-skin on his shoulder.
>
> Mary Louisa Whately, *Child-Life in Egypt* (1866)[30]

Among the infrastructural changes that had a direct impact on people's day-to-day lives was the introduction of water pipes and running water. As with the infrastructural changes discussed in the Introduction, the installation of plumbing was gradual and usually class-based: the wealthier and more modern areas of Egyptian towns and city had access to running water first, and the poorer and more rural areas did not have running water until much later. This transition also had a sonic component to it, since one of the most iconic sounds of Egypt's urban areas throughout the late nineteenth and early twentieth centuries was the call of the water seller, or *saqa* (pl. *saqayin*). To put the pervasiveness of these calls into perspective, by the late nineteenth century there were approximately ten thousand water sellers in Cairo alone.[31]

Before the mid-nineteenth century, when water plumbing was gradually introduced to some of Egypt's urban areas, water selling and distribution was a large and elaborate enterprise. Water merchants sold and transported large amounts of water by camel and donkey cart, filling large, centralized reservoirs and public baths, and essentially fulfilling all the water needs of Egyptian towns and villages. Plumbing and water piping gradually became more common in the second half of the nineteenth century among modern and wealthy households, which then had direct access to drinking water. However, Egypt's water economy dramatically changed soon after, as water companies and municipalities placed water fountains in central locations throughout large Egyptian towns and cities, providing easier water access for the retail water seller (and directly to some consumers), who would then transport that water to private residences and businesses. Most of these taps had an attendant who was responsible for opening the tap and distributing the water, for a small service fee. According to André Raymond, in Cairo alone, the municipal water system supplied 7.8 million cubic meters of water in 1882, and expanded sixfold to supply around 50 million cubic meters of water by 1927. The old part of the city had about 225 public hydrants for the residents who still did not have their own running water. Direct household accessibility to municipal water has continued to slowly increase in Cairo, reaching almost 78 percent of households by 1986.[32]

Until the majority of households had internal plumbing, which did not happen until well into the second half of the twentieth century, the *saqayin* continued to sell water to households after buying it from these centralized water taps and fountains. Their numbers noticeably decreased year after year. Unlike the vast majority of street merchants, the urban water sellers in Egypt were mostly regulated and part of a guild system that was hundreds of years old.[33] The *saqayin* crooned aloud the traditional chant "*Ya'awad Allah*" (God will compensate) to advertise their arrival to a neighborhood. The profit margin for transporting water was minimal, and it was expected that customers would tip the *saqa*, especially if he delivered the water to an apartment that was several floors up. So, the chant "God will compensate" was intended, in part, to remind the customer that a generous tip was needed and appreciated.

Aside from the *saqayin*, who predominantly sold relatively large quantities of water to fill up household and public basins and reservoirs, there were also retail water sellers, selling or often "giving away" glassfuls of water for a tip to passersby. These retail water sellers, like the sellers of lemonade and licorice drinks, usually advertised their presence not by vocalizing but by clinking two brass saucers together. For many tourists and travelers visiting Egypt at the time, both the traditional "*Ya'awad Allah*" cry and the distinctive clanking sounds of the brass plates, left a lasting impression. For example, Douglas Sladen, the British academic and travel writer who visited Cairo in 1911, remarks, "[T]he peripatetic water-seller is much in evidence here too, walking about with a gorgeous brass essence-bottle with a tapering spout, a water pitcher, and a couple of brass saucers, which are as tuneful as bells when he clinks them together."[34] References to the sounds of Egypt's water sellers abounded in most of the travel literature I examined.

Although in towns and cities, all of the "professional" water carriers and water sellers were men, in my assessment, the majority of the water consumed by Egyptians at that time was actually carried by women. For most of the twentieth century, the vast majority of Egyptians lived in the countryside, where it was mostly women who daily transported water directly from the Nile or its canals and tributaries to their homes and farms. In fact, even in the urban areas, where male water sellers were common, a large percentage of households could not afford to regularly pay the minimal amount that these *saqayin* charged, and it was primarily women who transported water for their own household use (see Figure 1.5). As an 1866 account of Bab al-Bahr district

FIGURE 1.5. Cairo: Women Drawing Water from the Nile.
SOURCE: "Femmes puisant de l'eau dans le Nil," EI-182, Bibliothèque nationale de France, Département des estampes et de la photographie, 1910, http://catalogue.bnf.fr/ark:/12148/cb44378144t.

in Cairo (near Azbakiyya and Bab al-Sha'riyya) relates, every afternoon, "the old men sit and sip their coffee opposite the coffee-house" and "the women crowd the street with heavy water pitchers on their heads."[35] This held true before the prevalence of municipal water taps and well after that, as many Cairene women living in the poorer districts then filled their water pots directly from these taps.

By the first quarter of the twentieth century, it was well understood that the entire *saqa* profession was becoming obsolete. As indoor plumbing and water faucets and pipes became pervasive in urban Egypt, the sounds of the water sellers' calls faded from the urban soundscape. Unlike street hawkers and other merchants who are to some extent still heard in the streets of Egypt today, by the mid-twentieth century, water sellers and their cries disappeared almost entirely from Egypt's urban areas.

Street Hawkers and Their Calls

> But the streets of all major European towns were seldom quiet in those days, for there were the constant voices of hawkers, street musicians and beggars. . . . The ubiquitous street cries were

> impossible to avoid. 'The uproar of the street sounded violently and hideously cacophonous,' reported Virginia Woolf in *Orlando*; but this is too general. Actually each hawker had an un-counterfeiting cry. More than the words, it was the musical motif and the inflection of the voice, passed in the trade from father to son, that gave the cue, blocks away, to the profession of the singer. In the days when shops were on wheels, ads were vocal displays.
>
> R. Murray Schafer, *The Soundscape*[36]

Aside from the water sellers, commuters, and carriage drivers already discussed, up to at least the middle of the twentieth century, a significant number of the Egyptian urban population, both men and women, made their living in the streets as beggars, street children, entertainers, fortune-tellers, donkey boys, pimps, prostitutes, taxi and tram drivers, and a dizzying number of street merchants and hawkers selling everything from vegetables to knickknacks and small trinkets. Indeed, street hawkers and their varied calls and sounds were, and to some extent still are, a vital part of Egypt's soundscape. As Murray Schafer describes in the section epigraph, with regard to early-twentieth-century American and European streets, the cries of the street hawkers were never just about loudness. What made Egyptian street hawkers successful at advertising their wares were the unique vocal inflections and musical tones reflected in those calls, which uniquely signaled to the customers precisely what was being sold.

Schafer assumes that most, if not all, of the European and American street hawkers were men. That was certainly not the case in Egypt, as a significant number of the street hawkers were women (see Figure 1.4). In fact, many late-nineteenth- and early-twentieth-century observers (and listeners) have noted a somewhat gendered division of labor among the male and female street hawkers. Mary Louisa Whately, an English missionary and educator who in the 1860s and '70s lived in the district of Bab al-Bahr in Cairo, remarked that Egyptian women typically sold milk, clotted cream, yogurt, oranges, radishes, grilled corn, sugarcane, and other fresh produce. Though she also remarked that she did "not recollect seeing a woman selling sweetmeats: probably it is considered a dignified business, as consisting of manufactured articles."[37] Muhammad 'Umar corroborates and expands on Whately's assessment in his 1902 book, *Hadir al-Misriyyin 'Aw Sirr Ta'akhurihum* (The present state of Egyptians, or, the state of their retrogression). According to 'Umar, "both men and women go around carrying their products for the entire day hoping to make a profit (a living)." 'Umar then specifies that men, on the one hand, typically specialize in selling "matches, books, shoes, sweets, clothing articles, pistachios, *batarikh* [bottarga: i.e., dried grey mullet roe], textiles, cheese, breadsticks, newspapers, peanuts, oils, and tamarind. They also sell whistles, bracelets, and everything for a penny [*kul haga bi 'irsh sagh*] and pottery." Women, on the other hand, sell "flowers, textiles, rosewater, dates, milk, honey, ghee . . . oranges, dates, and corn."[38] This observed division of labor between male and female street merchants was not hard and fast, as I have found plenty of evidence of women selling jewelry and, especially, farmer's cheese. In any case, there is no evidence of any official or social mechanism that specifically forbade female street hawkers from selling a particular product.

During the first half of the twentieth century, the number and variety of calls and cries from street merchants dwarfed what one hears today. An observer in mid-1930s Cairo documented 165 distinct calls and cries by street hawkers, and that was just for the sellers of produce and other food items.[39] The sellers loudly and melodically emphasized the freshness, ripeness, taste, size, or geographic origins of the foods they were selling. Comparisons, often amusing or exaggerated, were sometimes used to drive the point home to buyers. Tomatoes, for example, could be the size of pomegranates or as red as roses. Because Egyptians preferred small cucumbers, these were often likened to kidney beans. Some of the more unusual sale calls declared that the sellers' dates were as good as lamb. Pistachio sellers compared their product to roasted game birds, and sellers of barbequed corn announced that their corn was as tasty as roasted chicken. According to Fathi Radwan, in the Sayyida Zaynab district in Cairo, some sugarcane sellers called out "whole [i.e., in one piece] O sugarcane [*salim ya qasab*]," and grape sellers would alternate their calls between "like goose eggs" and "like pigeon eggs," depending on the size of their grapes.[40] If the tenderness of the food was important, then the sellers loudly chanted, "even the toothless can eat them." Upon seeing a child playing in the street, candy vendors would yell out "'*ayat li-ummak ya waad*," or "cry to your mother, little boy," urging the young child to beg his mother for money in order to buy candy.[41]

There were dozens of calls and also nonvocal auditory cues used by street vendors selling candy and other sweets to children. Remembering the variety of these distinct calls from his childhood during the late 1910s and early 1920s in Cairo's Sayyida Zaynab District, Fathi Radwan recounts that "it was as if each of those sweets was a person with a unique personality." For example, a young woman who walked the streets of Sayyida Zaynab carrying an "old rusty metal tray" sold *ali-luz*, which Radwan describes as a "sugar caramel candy covered in shaved almonds." The seller of a traditional chewing candy had the long pieces of toffee-like candy attached to a long pole with a rattle (*shukshikha*) at the end of it. According to Radwan, when the vendor shook the pole, "it produced a loud irresistible rattling sound that drove all the neighboring children crazy as they came out of the woodwork with their pennies [millimes] in hand." Radwan also recalls a cotton candy seller and a Turkish man selling *dondurma* (Turkish ice cream). A vendor selling vanilla wafers employed a variety of sonic tools to advertise his arrival. He not only "blew on a small horn" but made a loud "cracking sound by beating two pieces of wood against each other," and for good measure he sometimes shouted "vanilla cookies" (*baskut vaniliyyaa*).[42]

The Sounds of Metal

As mentioned earlier, many street vendors did not need to vocalize in order to sell their products but used a variety of other sounds to project their arrival to their customers. Butane tank merchants, who are still heard throughout Egypt, could signal their arrival to an entire neighborhood without uttering a single word. Striking their metal butane cylinders with a wrench, they produced a familiar loud and rhythmic clanking that sonically carried for several city blocks and signaled the arrival of the truck or horse-drawn cart. Like the retail (by the glass) water sellers described previously, lemonade, sherbet, iced hibiscus tea (*karkaday*), and licorice juice (*'ir'isus*) sellers, rhythmically clanked their large, castanet-like brass saucers to advertise their cold drinks. That clinking sound was most commonly heard during the summer. According to Mary Whately, in the hot summers of 1860s Cairo, the sounds of her neighborhood's sherbet seller predominated, as "the tinkle of his brass cups was sometimes the only sound in the hot and dusty street during the sultry afternoon hours."[43] Cold drink sellers and their clanking sounds remained prevalent in the traditional districts of Cairo well into the mid-twentieth century, and were often depicted by travelers in their photographs and travel accounts. Perhaps the earliest sounded film recording of one of these juice sellers was captured in the winter of 1928 by the Fox Movietone Company, in the newsreel described at the beginning of this chapter (see Figure 1.6). A licorice juice seller, surrounded by a mass of people in Cairo's al-Husayn district, performed in front of the camera, while clanking his brass saucers, for close to twenty seconds.[44] This iconic tinkling sound can still occasionally be heard in parts of old Cairo and elsewhere in Egypt on certain special occasions during Ramadan, Eid, and *mawalid* (or saint's day) nights.

FIGURE 1.6. Licorice Drink Seller and his Brass Saucers.
SOURCE: *Cairo street scenes—outtakes*, Fox Movietone News Story, December 28, 1928. (University of South Carolina, Moving Image Research Collections). Copyright holder: University of South Carolina.

The Temporality and Seasonality of Street Hawkers' Cries

> It is from the house-tops that the street criers, so characteristic of every nation, as showing the wants and tastes of the masses, are best seen and heard. Many of these vary with the season, as with us; but the one that begins the day never changes. . . . Just as the first ray of sunshine breaks forth, the muezzin's cry is heard, "To prayer, to prayer, O ye believers!" . . . Then, when the echoing voices from minaret to minaret have died away, the "working-day" begins. . . . First is heard the milk-woman's call, answering to "milk below!"
>
> Mary Louisa Whately, *Child-Life in Egypt* (1866)[45]

As I briefly mentioned in the Introduction, aside from the obvious new technological sounds that have drastically changed the soundscapes of the twentieth century, there is a constant seasonality and temporality to sound, which resonates differently depending on the time of the day and of the year. In this section's epigraph, Mary Whately captures the morning sounds and rhythms of most of Egypt's towns and cities: from the call to prayer that acts as a sort of sonic clock dividing up the day from dawn to after sunset, to the street hawkers beginning the day with selling breakfast items to their clientele. Whately notes that the milkwoman also sells yogurt for breakfast, and that "sometimes a woman passes bearing a tray on her head, with small earthen-ware bowls filled with the cream of buffalo's milk *scalded*; this dainty is called *Kishtar* [*qishtah* means clotted cream], and is much liked by Europeans as well as natives."[46] Other Egyptian foods like *ful* (stewed fava beans) and *bilila* (wheat porridge) are still primarily sold for breakfast, and the voices of the sellers of these foods and the sounds of their creaking wooden carts are still heard in the mornings throughout Egypt. According to yet another visitor to Cairo right before the turn of the twentieth century, one of the fava bean merchants yelled out in the mornings, "The beans of Embebeh [Imbaba] are better than almonds. Oh-h, how sweet are the little sons of the river!"[47] Referring to fava beans as almonds is common among street sellers in Egypt, as I recently heard a fava bean seller in Alexandria simply call out melodically and repeatedly *al-luz* (almonds) to advertise his beans.

Hawkers' cries did not change just over the course of the day but also according to the season. On hot summer days and nights, the number of cold drink sellers and the sounds of their cries and the clinking of their large brass saucers multiplied immensely (see Figure 1.6). Also, because the sun sets a few hours later, street activity continued much longer in the summer than in the winter. Seasonality also played a role in the available types of produce, which sets yet another, longer rhythm to the sights and sounds of the streets. Mary Whately captured the sonic seasonal rhythm of her Cairo neighborhood in a passage that is worth quoting at length:

> In the autumn, as soon as the harvest is commenced, the seller of parched corn is heard at intervals all day, and her store of young ears of corn roasted in their own husks is much in request.... The orange-crier was generally a woman; so was the seller of radishes. A little later in the day the sherbet-crier was heard: he had more custom in summer than winter, and the tinkle of his brass cups was sometimes the only sound in the hot and dusty street during the sultry afternoon hours.... When the seasons arrive, the various fruits and vegetables have each their crier. Then, when the real hot weather set in, about the middle of April, the cucumbers were in abundance, and eagerly devoured by all classes; and the cucumber seller had a very musical and lively cry.[48]

The seasons also played a role in the number and sounds of the thousands of guides, taxi and carriage drivers, and street hawkers catering to tourists. Cairo and Luxor were primarily winter destinations for European tourists visiting Egypt, where tourist activity came to a halt during the hot summer months. As William Lawrence Balls, a long-time British resident in Egypt observed in the early years of the twentieth century, "There is a Cairo of the 'season' and a Cairo of the summer. The one is restless as a Catherine wheel, pivoted on the great hotels, and the other is nearly as peaceful and subdued as a Trappist monastery." Moreover, "Opera Square in the winter is a dancing maze of traffic; taxis, cabs, and donkeys, itinerant vendors and snake charmers, dragomans, and tourists. On a summer afternoon it is a wide emptiness."[49] If Opera Square and the area immediately surrounding the Azbakiyya Garden—where most of Cairo's major hotels were located—were quieter in the summer, that was not the case for the rest of Cairo at that time, as Douglas Sladen elaborates in describing Ataba Square from the vantage point of a particular café:

> This café is at the point where the Musky [Muski Street] debouches.... Here, as soon as the weather grows hot, the ranks of the hawkers are swelled by an army of ice-cream sellers, lemonade-sellers, water-sellers, and sponge and loofah sellers. The clang of the trams and the water-sellers never ceases; the fly-whisk sellers are incessant. The Ataba-el-Khadra is a wonderfully busy place; there is a never-ending stream of tramways and native buses and native funerals, and people hurrying to the trams.[50]

Of course, the sonic seasonal equation in Alexandria was almost the exact opposite of Cairo's, as middle- and upper-class Cairenes and other tourists flocked to Alexandria and its beaches to escape the heat. Unlike the hotels in Cairo, Alexandria's hotels were usually overbooked in the summer, and the street hawkers followed the beachgoers, selling everything from lemonade and ice cream to *friska* (wafer cookies), clams, and *ritsa* (sea urchin).[51] From the late nineteenth century on, summer beach life in Alexandria, from June to September, was booming with the sounds of outdoor life, commerce, and entertainment.

Hawking Door to Door

> The most musical, perhaps, of all our street cries was that of the seller of parched peas [i.e., roasted chickpeas], and a certain little nut or seed which is much eaten by children, and, I suspect,

by grown people also. "O parched peas! O nuts of love!" was given in a really pretty, chant-like manner, and with a good voice.

Mary Louisa Whately, *Child-Life in Egypt* (1866)[52]

In twentieth-century Egypt, street hawkers did not sell their wares only near the market or to the coffee shop clients who often sat at outdoor tables on the sidewalks. As new institutions like theaters, movie theaters, and stations for new transportation vehicles like steamboats, trains, and tramways were introduced, street hawkers followed with their wares and services. A surprising variety of these hawkers had regular routes in most urban residential neighborhoods. The loudness and the pitch of their calls were essential for reaching inside the homes and apartments of the intended consumers. For street hawkers, sales came from more than just the words being sounded; it was the unique inflection, tone, and pitch of a particular seller's voice that signaled from quite a distance the exact product being sold. The astute listener could then approximate the distance and time of arrival of the salesperson—giving that customer ample time to walk down the stairs and conduct the transaction. For smaller items, customers did not even need to descend to the street for the transaction to take place. Most households had handy wicker baskets attached to long ropes in which they could send down cash from the windows or balconies of their apartments and bring up the change and the purchased goods. In these scenarios, both the buyers (who were predominantly women) and the sellers shouted the details of the transaction for all to hear.

Building walls could create a visual partition separating private from public, but most walls could hardly stop wanted or unwanted sound from traveling through. Windows, balconies, and doors enhanced this sonic penetration and served as sensory portals connecting residents with their neighbors and their neighborhood. There was little sound insulation in walls in early- to mid-twentieth-century houses, and residences in poorer district did not have glass windows and so lacked even that slight buffer against street sounds. Road noises and even loud conversations were easily heard inside houses and apartments. In most cases, that was not necessarily undesirable, as connecting with your neighbors and interacting with street hawkers was not only convenient but the expected norm. Windows, balconies, and rooftops had the potential to become liminal zones that were neither private nor public, especially when they were used as platforms for social and economic engagement.

The sounds and noises that people heard inside their private homes and that emanated from street hawkers were not limited to their calls. Their mere presence in the streets in significant numbers for most of the day created regular spheres of physical interaction and conversation with pedestrians, other vendors, storeowners, and even residents who had their windows, doors, or balconies open. The more one listens to these interactions, the more it becomes apparent that the split between public and private spheres was never as rigid as one might have thought from reading about the so-called Islamic city. For instance, Mary Whately often observed that she heard from her window her neighbors talking to each other and talking to some of the street merchants, either for a transaction or as mere conversation. Her neighbor, for example, who regularly sat on a reed mat selling sugarcane at the street corner in front of Whately's window, "talked incessantly to anyone who came within ear-shot, whether customer or not." Whately also vividly describes how in the late afternoons when the merchants were closing up their accounts, she would regularly hear "a few sharp bargainers trying to get oranges, beans, etc. at a lower rate than before, the clatter of tongues was quite astonishing."[53] Whately could clearly hear these conversations and see many of those interactions from her window or rooftop. Windows, doors, rooftops, and balconies more directly focused the auditory and visual interactions between those who were partly on the inside and yet engaged to varying degrees with their neighbors or street hawkers on the outside, providing liminal spaces of direct interaction that were neither entirely public nor private. As I examine in Chapter 3, a similar spatial and social liminality takes place in and around tramways as a space of social interaction.

Conclusion

As we literally focus in and listen to the changing sights and sounds of the Egyptian streets, we are forced to become more attuned to the street-level, everyday realities experienced daily by ordinary men and women. As revealed in this chapter, there was an unspoken significance and even temporality and seasonality to mundane street sounds, ranging from vehicular and pedestrian traffic to the sounds of merchants and street hawkers. By examining how city dwellers used the public streets and the variety of new places, spaces, and transportation technologies introduced at the turn of the century, we glimpse, or rather perceive, the everyday mundane challenges faced by ordinary people. Appropriations of the public city streets, squares, and spaces were also taking place through the significant numbers of men and women who were making a living by begging or by buying and selling goods and services in every nook and cranny of the city.

As I examine in Chapter 2, however, the streets and other public places were not merely neutral spaces for everyday people to embody and use. At times, public streets were spaces very much contested between the Egyptian (and colonial) authorities, whose instincts were almost always to regulate and control, and the majority of Egyptians, who simply wanted to use the streets on their own terms. Although the intended official function of the streets was orderly traveling and commuting, people often used them to make a living and to celebrate, morn, protest, entertain, and be entertained. The often embodied and loud appropriation of the streets by everyday people for their own purposes, and the state's inconsistent opposition to such endeavors, was at the heart of a continuing struggle over "ownership" of the streets. In the next chapter, I will examine some of these physical street contestations while simultaneously accounting for some of the "civilizing" discourses in the Egyptian press, which tended to marginalize most people who loudly made a living in the streets.

2

Silencing the Streets
Classism, Fear of the Crowd, and
Regulating Sounds and Bodies

> The denial of lower, coarse, vulgar, venal, servile—in a word, natural—enjoyment, which constitutes the sacred sphere of culture, implies an affirmation of the superiority of those who can be satisfied with the sublimated, refined, disinterested, gratuitous, distinguished pleasures forever closed to the profane. That is why art and cultural consumption are predisposed, consciously and deliberately or not, to fulfill a social function of legitimating social differences.
>
> Pierre Bourdieu, *Distinction*[1]

PIERRE BOURDIEU'S ANALYSIS of cultural taste and class formation can shed some light on some of the cultural and social dynamics taking place in the Egyptian streets in the first half of the twentieth century. This becomes much more apparent when examining the Egyptian streets using a sensory approach. A sensory classification of sorts, ranking Egyptian society along a deliberate yet subjective scale of vulgarity, is readily apparent in many of Egypt's twentieth-century cultural productions.[2] These classist discourses were a dominant theme in the Egyptian press and, as I examine in the rest of this chapter, were mostly a by-product of a small yet growing, upwardly mobile, educated Egyptian middle class.[3] Insecure perhaps in their newly acquired and still forming class position and armed with an almost messianic civilizing mission to uplift the Egyptian masses to some subjective and idealized level of "modernity," these writers and intellectuals vulgarized, infantilized, and sometimes dehumanized the vast majority of the Egyptian population. Much of this apparent disdain for the poor and those deemed uncultured was expressed in an explicitly sensory idiom that classified the masses as loud, dirty, tasteless, smelly, overtly sexualized, superstitious, and a danger to public health and public order. An important function of such categorizations, as Bourdieu demonstrates, is also to classify the classifiers, who are thus "distinguishing themselves by the distinctions they make."[4] But as I will elaborate in this chapter, this was not merely a matter of aesthetics, cultural taste, or reinforcement of class distinction but at times also a reflection of a deep-seated suspicion of the masses and a growing anxiety about the potential of street disorders.

These discourses had a direct and mostly detrimental impact on Egypt's poor and were translated into some oppressive state policies, leading to a recurring cycle of silencing and stifling all those dependent on the streets for survival. Indeed, the price of these often futile attempts at creating what Bourdieu describes as a "more polished, more polite, [and] better policed world" was and still is being paid in full by those who are most vulnerable.[5] Egypt's itinerant poor, from street hawkers, entertainers, and fortune-tellers to beggars and the homeless, were at the front line of these ideological and physical policing battles over control of the streets.

This chapter will, in part, document many of these often unsuccessful yet violent and disruptive attempts by the state to silence, "order," and control the streets. But more importantly, the varying ways which ordinary men and women accommodated, resisted, or coopted the state's increasing intrusions into their everyday lives will be a repeating feature of not only this chapter but also the remainder of the book. However, the panoptic powers of the modern state were never as neat, efficient, and controlling as they are sometimes portrayed. Everyday people daily used and "misused" the streets, for their own purposes, to loudly sell, buy, mourn, and celebrate, often breaking a host of state regulations in the process. Police enforcement, if it took place, was loud, messy, and contentious. If the enforcement was "successful," it was almost always resisted, negotiated, and at best partially or incompletely imposed. As I demonstrate in this chapter, by listening in to the sources, we can somewhat escape a totalizing, all-seeing paradigm, and more importantly for our purposes, we can attempt to uncover some of the lived experience of ordinary people and to reveal the changing sights and sounds of Egyptian street life.

Muhammad 'Umar and the Middle Class's Inferiority Complex

Muhammad 'Umar's *Hadir al-Misriyyin 'Aw Sirr Ta'akhurihum* (The present state of Egyptians, or, the state of their retrogression) is perhaps the most representative statement of the classist attitudes prevalent among Egypt's educated elite during the first half of the twentieth century. 'Umar, who was a middle-class government bureaucrat working at the Egyptian Post Office, gained quite a bit of notoriety when he first published his book in 1902.[6] 'Umar acknowledges in his introduction that his book was inspired after he read an Arabic translation of Edmond Demolins' *A quoi tient la supériorité des Anglo-Saxons* (Anglo-Saxon superiority: To what it is due).[7] The translator of Demolins' book was none other than Ahmad Fathi Zaghlul, the elder brother of Saad Zaghlul, the future Egyptian Prime Minister and nationalist leader. Unlike his brother Saad, Ahmad Fathi Zaghlul was notoriously anti-populist, and less than a decade later would become infamous for his role as one of the Egyptian judges in the Dinshaway trial.[8] Ahmad Fathi Zaghlul's introduction to his translation of Demolins' book applied some of Demolins' analysis to Egypt, casting the Egyptian masses as hopelessly ignorant and requiring a major civilizing uplift.[9] This no doubt provided the fertile seed for 'Umar's book, and not surprisingly, Zaghlul also wrote the preface for that book.[10]

Umar's classist, anti-populist approach in *Hadir al-Misriyyin 'Aw Sirr Ta'akhurihum* can only be described as an intolerant tirade against the Egyptian poor, whom he admits were essentially the vast majority of the Egyptian population. After a biting critique and an exaggerated stereotyping of Egypt's traditional wealthy elites, he somewhat sings the praises and elaborates on the civilizing potential of the small yet growing, upwardly mobile middle classes: of which he of course counts himself as a true representative. In the last third of his book, 'Umar lambasts the Egyptian poor, as thieving, ignorant, and backward with almost no redeeming qualities.[11] 'Umar, as we will soon see, has a lot to say about street hawkers and the itinerant poor.

Though 'Umar was extreme in his views regarding the poor, he represented a classist "civilizing" attitude widely shared by many Egyptian intellectuals throughout the twentieth century. In Zachary Lockman's detailed analysis of 'Umar's book, he correctly places him squarely in the "reformist" Muhammad 'Abduh camp, "operating more or less within the same discursive field as such contemporaries as Ahmad Fathi Zaghlul, Yusuf al-Nahhas, Qasim Amin, and Ahmad Lutfi al-Sayyid."[12] Some of these classist attitudes with regard to the uplifting of the ignorant and unwashed Egyptian masses would be represented in the Umma Party's platform and in Lutfi al-Sayyid's *al-Jarida* newspaper when these vehicles were created in 1907.

Many of the reformist "concerns" were (and still are) especially directed toward the physical public presence of the itinerant poor in the streets, squares, and alleyways of Egypt's towns and cities. Discourses emanating from the Egyptian and colonial press were at best condescending and at worst vulgarizing toward all those who lived or made a living in the streets. This fear of the hubbub of the "lower" classes characterized not just the discourses of the "upper" classes and the cultural elites but was also reflected in the writings of the journalists, editors, and intellectuals who filled the press with a steady stream of articles, opinions pieces, and exposés about the urban poor and all those who lived on the margins. Indeed, for many in Egypt's growing middle classes, the urban poor became a sort of national shame that needed immediate civilizing and uplifting, reminding the middle classes of how far behind the country was in relation to their utopian modernizing vision. Pickpocketing, petty crime, begging, prostitution, disorderly conduct, and various other street swindles regularly filled many daily and weekly periodicals. The Akhbar al-Hawadith (accidents and crime) sections of various newspapers and regular letters to the editor recorded often exaggerated reports of the various threats emanating from the streets and street people. As is often the case, these alarmist representations of the masses almost always utilized demeaning sensory vulgarization. These reports typically portrayed Egyptian masses as loud, boisterous, smelly, tasteless, and uncouth. Quieting or at least muffling the sounds of the masses was always presented as the first step in civilizing these future citizens.[13]

The "Latent Deviance and Immorality" of Male and Female Street Hawkers

Writing at the turn of the twentieth century, Muhammad 'Umar was characteristically judgmental of street hawkers in general and was not just critical of their commercial activities in the streets but also always suspicious of their "social" interactions with customers and passersby. Like many of his contemporaries, he seemed to be always fearful of what he describes as the chaos of the streets, and the lower classes were almost always the instruments of that chaos.[14] After a long soliloquy decrying the laziness and ignorance of the poor, 'Umar begins to describe some of the "unproductive work" they engage in, including peddling and street hawking. Street hawkers, according to 'Umar, are all thieves and swindlers at heart, who either misrepresent what they sell or manipulate their scales by "cheating with the weight and measure of what they sell." He emphatically declares that "both men and women are guilty of this offense."[15] He seems to especially detest female street hawkers, fearful of what he saw as their latent sexuality, and practically accuses all of them of borderline prostitution:

As for the girls that start out as sellers of oranges or dates we can say that they start out selling these things while maintaining some respect and covering their faces when approached by men because of respectability and shyness. After a bit of time passes, these girls then abandon the hijab entirely and walk joyfully without it. They later learn the latest jokes and jests (*hizar*), and when a passer-by, donkey boy, or a carriage driver passes by, these women flirt and joke with them, and they even dare to do this with some policemen. I recall one day as I was standing near one of these women, a policeman approached her as she was sitting on the ground [immodestly] with her legs wide open. They were both laughing out loud as they loudly shared jokes and conversed familiarly with each other. When it was time for him [the policeman] to leave, he took some of what she was selling and made sure to put in a good word for her to the next policeman taking over his shift. This is the sort of thing that typically happens with the rest of these female street sellers.[16]

What was most likely just friendly banter, was viewed by the conservatively repressed 'Umar as public sexual flirtation, enhancing in his mind the potential moral and sexual chaos that a poor woman working in the public streets represents. There is another issue, however, implied in 'Umar's account. As he left at the end of his shift, the police officer took some oranges from the woman without paying for them. If this was the case, then the street vendor was most likely trying to guarantee that the officer would allow her to continue to sell her oranges in this location. According to 'Umar's account, this interpretation was corroborated by the fact that he overheard the officer "putting in a good word" for the orange seller with the next officer on shift at the same location. Regardless of the exact veracity of this account, it is certainly plausible, since there were, on paper at least, a few laws either regulating or in some locations forbidding street hawking altogether. In any case, the relationships of the police and street hawkers were not always as "accommodating" as in the instance described by 'Umar. As the rest of the chapter will show, there were certainly many documented cases of police abuse and mistreatment of street merchants and the itinerant poor.[17]

Regulating and Quieting Street Hawkers

In a curious and rather unrealistic attempt at profit making, an enterprising businessman named M. G. Eram saw the untapped financial potential of taxing the thousands of street hawkers in Cairo and Alexandria. On January 12, 1889, Eram petitioned the Egyptian government to license him and give him a personal concession to become the sole tax-collecting agent responsible for tracking down Cairo's and Alexandria's street merchants and soliciting taxes from them. In making his case to the Egyptian government, Eram claimed that 80 percent of all street merchants were outside the jurisdiction of the municipal authorities and were "not known to the leaders of the city quarters (*mashayikh al-harat*) and the heads of the guilds (*mashayikh al-hiraf*)." To tempt government officials to agree to his scheme, he reminded them of the potential tax revenues that could be gained if the street hawkers were registered and taxed. Eram then proposed that he be given a monopoly on collecting taxes directly from the street merchants of Cairo and Alexandria for a period of two years, and in return he would pay the state 5,000 Eqyptian pounds (LE) per year.[18] This concession, which sounds more like an organized crime racket than a legitimate business venture, was of course never implemented, and it is almost certain that this scheme would have failed had the Egyptian authorities allowed Eram to execute it. Without an enforcement mechanism to implement the taxes on the street hawkers, it would have been impossible for Eram to recoup his investment.

However, Eram correctly assessed the existence of thousands of mostly unregulated street hawkers in turn of the twentieth-century Egyptian towns and cities. The lack of a regular fixed address and the ability to move virtually anywhere provided hawkers with a great deal of flexibility and the potential to evade taxation and regulation. Despite newer, more advanced surveillance techniques employed by an increasingly centralized Egyptian (and colonial) state, most street merchants could still contest, resist, and at times ignore the state's attempts to control their trade and their bodies. Overall, enforcement of these laws was limited and inconsistent, although it could quickly turn capricious and arbitrary at times of political turmoil.

Ideally, of course, the state wanted to consistently control the streets and especially hawkers and merchants, though its attempts were almost always an exercise in futility. Some level of control could be achieved if the state had the will and the resources, and if hawkers had regular routes through the city, but we can safely assume that the majority of street vendors and hawkers, whether in the nineteenth century or today, were well able to evade regular control by the state. However, because the hawkers' marginalized political and economic status often left them on the margins of society, individual police authorities irregularly and arbitrarily harassed and abused them for bribes and kickbacks. The Egyptian archives are full of petitions by street hawkers and vendors complaining of constant police abuse, corruption, and harassment.

In the twentieth century, monitoring and dictating the behavior of street hawkers by means of laws and regulations began in earnest on January 31, 1915, with a directive from the Minister of the Interior. According to the letter of this law, each hawker needed to acquire a license that had to be renewed on an annual basis. The number of this license was supposed to be on display. Aside from paying the fees, each hawker had to be cleared medically and had to have no criminal record. Since all street vendors loudly advertised their wares, their hours of operation were limited by law. It is of course doubtful that these regulations were fully enforced. In fact, in 1941, a much more robust law was passed and at least partly implemented. An explanatory memo included in the 1941 government file explicitly stated that the 1915 law was ineffective and insufficient.[19]

The 1941 law was based on the recommendations of the Egyptian Prime Minister and the Ministry of Public Health and was presented to the Egyptian parliament by the office of King Faruq at the end of July 1941. It was, in effect, an updated revision of prior legislation regulating street hawkers, essentially expanding upon and amending the 1915 law. The 1941 law consisted of thirteen articles dictating almost every aspect of the lives of street vendors.[20] The law defined a street hawker fairly broadly as anyone who sells goods while moving around in the public streets, whether by carrying the products on their person or by using a cart or any other vehicle. According to Article 2 of the law, a street hawker must acquire a yearly license from the nearest governorate or district. The license took the form of a sequentially numbered passport-like booklet, costing 30 millimes per year.[21] For an additional 50 millimes, the hawker received a metal license plate engraved with the number and the name of the issuing governorate or directorate. The metal plate was supposed to be displayed on the street hawker's left arm at all times, and the paper license was to be shown to the police or to the health authorities upon request. In order to receive a license, a street hawker had to be at least twelve years old, demonstrate that he or she had not committed a crime in the last two years, and document that they were free from any skin or intestinal disease. The government reserved the right to revoke the license if the licensee committed a crime or became ill with a skin or intestinal disorder. A person whose license had been revoked owing to an illness could get the license restored by acquiring a medical certificate endorsed by the proper health authorities and indicating a clean bill of health. Under the letter of the law, anyone who was arrested for selling anything in the streets without a license or who broke any of the health laws or the rulings passed by the Ministry of Public Health could be jailed for up to one week and fined up to 100 piasters. As we shall soon see though, this was very irregularly enforced.

Siestas and Silencing Street Cries

Street hawker regulations not only dictated where and when these vendors could sell their wares but also attempted to control their hygiene, the smells and freshness of their food, and the volume of their verbal and nonverbal calls. Article 9, Section 3 of the 1941 law, for example, explicitly forbade street hawkers from "pursuing the public or advertising their wares by yelling, ringing bells, blowing horns or any other similar instruments, in a fashion that disturbs public peace and tranquility." The regulations even attempt to dictate the times during the day when hawkers could make their sale rounds. For example, on paper at least, during the winter months, street hawkers were not allowed to make their calls between the hours of 8:00 p.m. and 7:00 a.m. During the summer, they were forbidden to sell their wares from 10:00 p.m. to 6:00 a.m. Interestingly, from April 15 to October 14, they were also forbidden to sell anything during siesta time, from 2:00 p.m. to 4:00 p.m., so as not to disturb those who were napping in the hot summer afternoons. Article 9, which explicitly lays out things that street hawkers are forbidden to do has six sections, and it is worth quoting in its entirety:

> Article 9
>
> Street Hawkers are forbidden from:
>
> Section 1. Selling their goods or walking through and standing in any streets or Square forbidden to Street Hawkers as directed by any written decree by a Governor or Director [*Mudir*].
>
> Section 2. Standing close to, or selling their goods near any store which sells similar items. Street Hawkers are also forbidden from standing or selling their goods in any place forbidden by the police in order to facilitate traffic, or maintaining order and security.
>
> Section 3. Pursuing the public or advertising their wares by yelling, ringing bells, blowing horns or any other similar instruments, in a fashion that disturbs the tranquility of the public.
>
> Section 4. Selling food or drink that is spoiled or banned by the Department of Public Health.
>
> Section 5. Selling of explosives, weapons or fireworks.
>
> Section 6. Pursuing their trade between the hours of 10:00 p.m. and 6:00 a.m. from April 15th to October 14, and between the hours of 8:00 p.m. and 7:00 a.m. during the period of October 15 to April 14th. They are entirely forbidden from advertising their wares by calling out or by any other means from 2:00 p.m. to 4:00 p.m. in the period of April 15th to October 14th.[22]

There was a great deal of ambiguity in these laws, and they were certainly open to interpretation and misinterpretation, which caused much hardship for many who relied on the street for their livelihood. Article 10 in particular gave a great deal of power to the governors and the *mudiriyya* directors. They had the right to deny licenses for a particular area or to limit the number of licenses issued. They also apparently had the power to exempt certain types of street hawkers from this law, with the exception of hawkers who sold food and drink, who still had to acquire a permit from the Ministry of Public Health. Petitions made to the Egyptian government by street hawkers and some of their advocates seem to indicate that licenses were limited in number and also very difficult to obtain, if they were available at all. If a street vending license was available, there were still quite few bureaucratic hurdles that had to be passed before it was acquired, not the least of which was following through with all of the procedures and regulations put into place by the Minister of Public Health. This, at best, caused confusion and, at worst, opened the door to corrupt practices and bribery of the police and other government officials.

Resisting and Petitioning the State

The vast majority of street hawkers were unregulated and unlicensed, and though extensive laws and regulations were on the books, they were unevenly enforced. Also, just as today, police abuse of street merchants was common.[23] As dozens of petitions to the Egyptian King and the Egyptian government in the 1930s and 1940s attest, arrests, police beatings, and the embezzlement of money from unlicensed merchants regularly took place (see Figure 2.1). The most common reason for the arrest of street hawkers was obstructing traffic. The arresting police officers would invoke the second section of Article 9, which states that "street hawkers are prohibited from standing or selling their goods in any place forbidden by the police in order to facilitate traffic, or maintaining order and security." This was by far the most commonly abused clause in the law as it was wide open to interpretation, and it allowed the police to arrest street hawkers on any whim.

In an October 16, 1944, petition to King Faruq, Abd al-Maqsud Khalaf Hasan, the self-proclaimed president of the Public Union of Knick-Knack Peddlers (al-Niqaba al-'Amah li-Ba'i'i al-Khardawat) wrote:

> Your Majesty,
> We are petitioning you because of the mistreatment we suffer at the hands of the police, for they are constantly harassing and chasing after us in the streets and al-

leyways as if we are criminals running from justice. The truth is that we are merely trying to make an honest living.... Lately, the police have especially been hounding us and upon arresting a poor merchant, they simply accuse him of being a "street hawker." At the police station, the prosecutor typically writes up the case against the vendor and sets a court date, and a bail of one Egyptian pound.

If the vendor cannot afford to pay the bail they are imprisoned for up to eleven days [or until the court date]. How is a street merchant supposed to feed his family of 3, 4, or 5 individuals if he is in prison? And how is he supposed to pay that one pound, which is nearly his entire capital? One pound is almost the cost of the goods that he is carrying on his person. If he does pay that one pound, then most likely he will resort to becoming a criminal or swindler in order to feed his children. So, in short, an otherwise obedient law-abiding merchant is transformed into a criminal and a potential danger to law and order.[24]

FIGURE 2.1. Confrontation between Police and Street Merchants in Port Said (1900).
SOURCE: The author's personal collection of photographs (E. W. Kelley Publisher, Stereo View Image, 1900).

After declaring that street hawkers "do not have any rights," Hassan stresses yet again that street vending is the only line of "honest work" that is open to the hawkers, and as such it is their only means of support (see Figure 2.2). Street hawkers, Hassan elaborates, have to work out in the elements, enduring "the coldest of winter nights and the hottest of summer days." He then continues to seek sympathy for his cause, by describing the typical poverty experienced by most street hawkers, highlighting that despite enduring daily toil and backbreaking work, they can barely make ends meet. Stressing the dire economic situation during the Second World War, Hassan specifically discusses inflation and the dramatic increases in the prices of everyday staples. He then adds that all these hardships have been compounded by the police, who lately have been unfairly chasing after and arresting peddlers in the streets. Responding to the allegation of obstructing traffic, which was arguably the most common charge leveled by the police against street hawkers, Hassan declares: "Most of the products we sell are light western made goods amounting to a total of 2–3 kilograms at the most. Eastern knick-knacks, sunglasses, pens, and other writing implements, which are all very small and light; we certainly do not cause any traffic obstructions as is often claimed by the police." After making his case, Hassan then proceeds to lay out his four basic requests from the government:

These are our basic demands:

First: A generous order from your highness to all the necessary authorities in order to halt the present policy of chasing after and arresting street hawkers.

Second: Introduce laws concerning street hawkers that are compatible with our current modern age.

Third: Facilitating for us the attainment of street hawker licenses (in the same style as those available to newspaper sellers). This is with the understanding that we are willing to pay for the necessary fees and obey all regulations. This we will do in order to protect ourselves from the police.

Fourth: In order to protect public order from all evildoers and criminals, this union has decided to voluntarily submit all of its members to all the necessary background checks.[25]

Hassan and the other petition signers appear to be fully aware of the state's public security anxieties, and the petition cleverly allays the authorities' concerns about street disorders and criminal elements by volunteering to submit background checks for all union members. And interestingly, Hassan reverses the civilizational discourse emanating from the Egyptian press with regard to street vendors

FIGURE 2.2. Simit Bread Seller (ca. 1955).
SOURCE: Eugene Harris Collection, American Geographical Society Library, University of Wisconsin-Milwaukee Libraries.

and the urban poor by attempting to shame the police and the state into enacting enforcement policies that are "compatible with our current modern age."

Not surprisingly, none of these petitioned demands were met. Three years later, a dozen Cairene street merchants appointed a man named Selim Nicola, a resident of the Muski District of Cairo, to petition the King on their behalf. His petition, dated March 27, 1947, complained of the exact same police abuses and arbitrary arrests. As Hassan did in the previous petition, Nicola claims that "policemen were unfairly arresting, beating, fining, and detaining" street vendors. Nicola also suggests that the government license the street hawkers in order to legitimate their trade and prevent police abuse.[26] These two petitions, and others like them, suggest that the elaborate 1915 and 1941 laws concerning street hawkers were only partly enforced. So even though according to state regulations at the time, street merchants were supposed to have licenses, few of those licenses were given out, or they were very difficult to come by. In any case, these uneven efforts by the centralized authorities and the police to restrict the public movement and presence of street hawkers also extended to the poor, street children, and the homeless.

Policing the Streets and Silencing the Poor

> Since I became the Mayor of Cairo I received many complaints from the citizens of the capital who come from a variety of different nationalities. In these complaints they bemoan the social condition of the city and the biggest criticism was about the increase in homelessness which ruins the beauty and reputation of the city and greatly upsets the public. I paid special attention to this issue.
>
> 'Abd al-Salam al-Shadhli Pasha, Cairo Mayor,
> "His Highness the King Orders the Rescue of Homeless Kids—
> 600 Homeless Boys Are Sent to an Agricultural Encampment:
> An interview with the Honorable Mayor of Cairo" (1938)[27]

For most of the twentieth century, not only were the authorities and the Egyptian "public" concerned about street hawkers, but more broadly, many Egyptians were also apprehensive about the urban poor and especially the homeless. The line between a beggar and a street hawker often blurred. This was also the case for a dizzying variety of street entertainers and for the men, women, and children selling flowers, jasmine, and trinkets. The category of the homeless was quite broad and included beggars, street children, and those who were temporarily unemployed. It also included many who worked in a variety of "street jobs," from shoeshine boys to fortune-tellers, and who simply did not make enough money to afford regular lodging. Fortune-tellers, usually women, provided "psychic" readings, though there was an implicit charitable component to the transaction. Female Quran reciters were also a common sight in Egypt's streets well into the first quarter of the twentieth century. The usually bad-tempered Muhammad 'Umar had a great deal to say about this:

> Many poverty-stricken girls are seen daily by passers-by in the streets of the capital and in other cities, sitting down in the streets reciting the Quran for all to see and hear. They especially sit near al-Sayyida Zaynab and al-Sayyida Nafisa Mosques and the Mitwali and Abu- al-'Ila bridges. It is no secret that what they do is a grave disgrace

for us all, since they read the glorious Quran sitting in the dirt and mud while holding their children who are often screaming and crying, mixing in our ears the recitation with the sounds of weeping babies. It's as if the babies can feel that this is not lawful, so they cry proclaiming their innocence from this entire affair. Otherwise there is no need to cry for they say that children enjoy a good tune and a pleasant voice.[28]

Yet in this case, the incessantly judgmental 'Umar found a thread of potential redemption for the Quran reciters, but only if they were "taught and trained to read the Quran properly." Then, according to 'Umar, they could "make a proper lawful [*halal*] living as they correctly recite Quranic verses at funerals, this way they can be blessed by God as well and not cursed as these street women are."[29] Public female Quran reciters were commonly heard in the Egyptian streets in the early twentieth century, but they seemed to have disappeared by the second half of the twentieth century.[30]

The Egyptian press commonly printed new anti-poor, anti-begging, and anti-homelessness discourses invoking fears of an imminent breakdown of public security, public order, and even public health. Some even described beggars as "foci of infection and called for their removal from public spaces." As documented by Mine Ener, these discourses began as early as the middle of the nineteenth century, and I would argue, still continue today.[31] The Egyptian press was full of editorials and reports warning of the dangers of the streets and of street people, which had an effect on public opinion, as corroborated by the comments from Cairo's Mayor in the epigraph for this section. Often these classist articles had alarmist headlines as they talked about pickpockets, beggars, gypsies [*ghagar*], street children, or the homeless. As the cartoon in Figure 2.3 illustrates, the concern was not merely about the safety of the public but also about the "civilizational" appearance of the city streets. Printed on the cover of the December 10, 1934, issue of *al-Ithnayn* magazine, the cartoon dramatically exemplifies the insensitivities of these classist discourses. The clearly poor, Upper Egyptian peasant family is dehumanized and depicted as resembling the chimpanzees in the Cairo zoo. Leaving no doubt as to the malicious intent of the artist's representation, the father speaks to the chimpanzees, asking them, "Why don't you come out [of the cage]?" One of the chimps responds, "Why don't you come in here?"[32] Although this was an extreme example of the literal dehumanization of the working poor and peasants in the Egyptian press, it is nonetheless representative of the constant attacks marginalizing Egyptian subalterns. These articles and

FIGURE 2.3. Egyptian Peasants Caricatured as Chimpanzees.
SOURCE: al-Ithnayn, December 10, 1934.

cartoons were frequently printed during this period. Surveying *al-Ahram* newspaper in the 1930s and 1940s, I found only a few articles that defended the street poor; most tended to blame them for being lazy or accused them of criminality.

Starting in the mid-nineteenth century, there was a repeating obsession among the Egyptian and later on British authorities with clearing the streets of those who were considered undesirable. This included beggars, street children, the homeless, gypsies, street entertainers, and, as examined earlier, unregulated street hawkers. Typically, if arrested, able-bodied males begging or living in the streets were fined for their first offense and faced imprisonment as repeat offenders.[33] It is hard to fathom, however, how these men and women could afford to pay the fines, and it is likely that most faced imprisonment upon arrest. These laws were at best irregularly implemented, but when they were enforced, they were devastating to these marginal groups who depended on the streets for their very survival.

Vagrancy Laws and Blaming the Victims

> While not wishing to dwell too much on the volume of work of the *qisms* [police stations], I will quote the following figures from one *qism*, 'Abdin, to show what a maximum effort is required to obtain a minimum result in checking the daily nuisances of the town. During the year 535 contraventions were made for vagabondage, 448 contraventions were made for beggars, 2,398 contraventions were made for hawkers, and 731 contraventions were made for illegal tram-riding.
>
> *Ministry of the Interior Report for the Year 1922*[34]

The rounding up of urban beggars, "vagrants," and vagabonds by the Egyptian government goes back to at least the reign of Muhammad Ali (r. 1805–1848). Concerns for order and public health were always cited as reasons to remove the itinerant poor. There were also some unevenly enforced anti-begging and homelessness laws in the early 1890s.[35] And begging was outlawed, on paper at least, in the districts of Azbakiyya, 'Abdin, and Muski.[36] In the twentieth century, vagrancy laws in Egypt were revised and irregularly enforced in 1909, during an era of political turmoil and street violence culminating in the reapplication of the censorship press laws of 1881, and the eventual assassination of Prime Minister Boutros Ghali Pasha in 1910.[37] For this reason, and in order to curb any street political activities, the laws were fairly broad and left quite a bit of room for interpretation. The 1909 laws were revised in June of 1923, during the ministry of Yahya Ibrahim, and according to contemporary legal critics, they were made even more complicated and vague, opening up the door for police and judicial abuse.[38]

The revised 1923 law was titled, "The law on vagrants and on those under surveillance and wanted by the police," and contained thirty-five rambling articles detailing differing issues including probation, fines, and sentencing.[39] The two most relevant to our discussion are Articles 1 and 31. The thirty-first article specifies that the "law does not apply to women or children under 15 years of age." This is pertinent to the analysis later in the chapter. The first article contains seven sections, and was the most controversial because of its inconsistency and its vagueness. It is worth quoting in its entirety:

Article 1

Who is to be defined as a vagrant under the law?

Section 1. Those with no lawful means of support.

Section 2. Those who make a living through gambling in the streets or those who read people's fortune in the streets or in any other public place or in any place where they are exposed to the gaze of the public.

Section 3. Pimps [*quwad al-nisa' al-'umumiat*].

Section 4. Those who are capable of work and yet take to begging in the public streets.

Section 5. Those who were sentenced more than twice for forcing children to beg in the streets and other public places and who were last sentenced within one year.

Section 6. Gypsies that travel the country without having a permanent home, unless they show that they have a profession or lawful employment.

Section 7. Those who regularly spend the night in the public streets and squares of towns or cities and those that do not have a home.[40]

In 1933, a conscientious lawyer named Abdallah Husayn wrote a series of articles for *al-Ahram* on amending and improving various aspects of Egyptian law. One of these articles was titled "Tanqih Qanun al-Mutasharidun" (Revising the vagrancy law).[41] In this long essay, Husayn scathingly attacks the vagrancy law of 1923 and calls for it to be completely rewritten. He elaborates on both the law's unfairness and its vagueness, which he claims has led to many false arrests. More importantly to Husayn, the law allows for police abuse and is not equitable when considering people's circumstances, the poverty that they endure, and the desperation that stems from their hardship and their destitution. The vagueness of the law, according to Husayn, results in a wide range of inconsistent legal verdicts and leaves the door open for abuse. Apparently, politicians and the police, in times of political turmoil, gained a tremendous amount of leverage by misapplying the law to arrest political demonstrators. According to Husayn, there were unsuccessful attempts at "revising the law in 1927 during the ministry of Tharwat Pacha, and the idea was still in circulation in the Interior Ministry."

To prove how arbitrary some of the cases were, and to demonstrate the ambiguities of this law, Husayn cites a case where the police arrested some shoeshine boys by following the letter of the law and claiming that because they did not have a license they lacked the "lawful means of support" required by Article 1. At their trial, the judge sided with the defendants and overthrew the case, stating that shoe shining is a lawful means of employment even though the defendants did not have the proper license. Husayn makes a strong case for judicial reform and concludes that "poverty is not cured by punishing and criminalizing the poor."

There were certainly traditional, as well as newer, publicly and privately funded institutions that helped with feeding and housing Egypt's growing homeless population. But in times of economic strain, which included most of the 1930s, these institutions struggled to keep up.[42] Mosques, like al-Azhar, al-Husayn, and others throughout Egypt, provided places of last resort for many of the urban poor, yet they were not enough and did not provide regularly funded meals and proper long-term shelter. Some private and public institutions could temporarily house street children, women, and the old and handicapped. Furthermore, the Egyptian elites publicly supported charities that supported street children.[43] However, according to Mine Ener, most of the extensive plans to "help" and train the poor and especially street children, "remained but blueprints in terms of their successful enforcement," and "despite their intended goal of removing potentially criminal boys and girls from the streets of Cairo and Alexandria, juvenile 'vagrants' remained a public presence."[44]

On the official level, charity was given to the poor in established institutions like hospitals, orphanages, and reformatories, but increasingly the poor were not tolerated in the streets. In the Egyptian press, anti-poor discourses intensified at the turn of the twentieth century and continued well into the first half of the century, and this anti-poor sentiment was reflected in Egyptian laws and government actions. Anti-poor and anti-begging discourses not only cited fears of criminal activity and the physical danger presented by street people but also invoked fears of lack of cleanliness, disease, and contagion. In a presentation to the Egyptian Royal Society, Kamel Greiss broadly accused beggars of being "lazy peasants pursuing a career of begging instead of more difficult work in the countryside." These sentiments were apparently a reoccurring theme in other discussions and lectures at the Egyptian Royal Society.[45]

The "Undeserving" Street Beggars

One of the biggest dilemmas discussed in the Egyptian press was determining which people ought to be classified as the "deserving poor," and which ought to be classified as criminals or lazy imposters who feigned poverty and destitution. Some broad criteria of deservedness were created by the Egyptian government as new, specialized orphanages and state hospitals were created to take care of abandoned children and the poor who were in need of medical care. However, these criteria were insufficient, as many people simply fell through the cracks, and there were not nearly enough beds and rooms to put a dent in the overall problem. The one consistency in these discourses was an increasing concern about the appearance of Egypt's public spaces.[46] To the elites and many of Egypt's upwardly mobile middle classes, the mere existence of the public poor was a blemish on the face of the new, modern, and "civilized" Egypt that they wanted to portray. It is no wonder that few of the many urban poor depicted in the press were deemed as deserving of charity. The cartoon shown in Figure 2.4, published in 1935 in the popular entertainment magazine *al-Ithnayn* and titled "Is He Blind or Is He Pretending to Be Blind?", is a classic example of the typical representation of the so-called undeserving beggar. The cartoon's caption, written in vernacular Egyptian Arabic, reads:

> The Alms Giver: O damn, I dropped the five piasters on the ground!
> The Beggar: It's right next to my walking stick, but I can't bend down to pick it up because I can't see it![47]

In another example, if we were to believe the editors of *al-Ahram* newspaper, a twenty-year-old man named Ahmad Tawfiq belonged squarely in the imposter and undeserving camp. The January 6, 1934, *al-Ahram* article, uncreatively titled "A Homeless Man Claims to Be a Sufi Saint," sets its tone at the

FIGURE 2.4. "Is He Blind or Is He Pretending to Be Blind?"
SOURCE: *al-Ithnayn*, May 20, 1935.

beginning by stating that "every day the police arrest a large number of vagrants from all over Egypt and especially Cairo in accordance with the vagrancy law."[48] According to the reporter, Tawfiq—who was fingerprinted and placed under arrest by the police—was disheveled, wearing torn clothes made of a rough material, and "speaking incoherently as if he was drunk." The reporter then sarcastically adds that Tawfiq "claims to be a Sufi Saint and best friends with al-Sayyid al-Badawi, the one with the well-known mosque and Sufi shrine named after him in Tanta." By mockingly stressing that al-Badawi, who passed away in Tanta over 640 years ago, was "best friends" with Tawfiq, the journalist purposely biases the reader against Tawfiq and his pending case. It is important to note here that aside from the anti-begging discourse, the article also indirectly mocks Sufism, which was consistent with the disposition of the Egyptian media at the time. As I examine in more detail in later chapters, unofficial Islam, or Islam as practiced in the streets by everyday people, was generally frowned upon as regressive, and it was almost always portrayed as tainted with ignorance and folk naïveté.

So as to not leave any doubt as to the guilt and falseness of Tawfiq's claims, the reporter maintains that "because of this vagrancy law, many beggars come up with inventive ways to avoid arrest," and that one of those ways was pretending to be a Sufi saint (*wali*). Of course, by suggesting that this homeless man was a fake and pretending to be a saint, the article automatically biases the reader, and in the process gives the ultimate legitimacy to the state and whatever actions it intends to pursue against Tawfiq. What is truly tragic in this particular case, and surely in many others like it, was that it would seem that neither the police nor the reporter suspected that the young man might be mentally ill and likely a schizophrenic. The immediate assumption was that he was drunk, and pretending to be a Sufi saint either because it would justify his begging or because he was hiding from the authorities owing to some past crime.

In the case of Tawfiq, the trail ends here, as there is no way of knowing what eventually happened to him. However, this storyline and the way it was framed and represented in *al-Ahram* and in the cartoon in *al-Ithnayn* were characteristic of an entire genre of articles, photographs, cartoons, and other media that were unambiguously dismissive, if not outright hostile, toward beggars and street people in general. Panhandlers and drifters were almost always depicted as feigning poverty in order to make a profit. Very rarely were they humanized in the press as people who genuinely needed assistance. As far as most of the Egyptian press was concerned, the majority of beggars and street people belonged squarely in the "undeserving" and "imposter" camp. Such regular representations must have had an impact in justifying yet more repressive policies vis-à-vis the itinerant poor.

Reforming Street Children

> Wherever you go you see a bunch of these kids spread out everywhere in shabby clothes fighting over bits of food and bread or a cigarette butt. They are often sitting on the ground as if they have no mothers or fathers. They fill the streets and alleyways with loud screams and swear out loud as they have no education or proper upbringing. . . . You find these fearless boys and girls aggressively and cunningly attempting to pilfer or scam passers-by, stealing what they can get their hands on. They are forced to do such things because of hunger and destitution and they are not to blame for this but it's entirely the fault of their mothers and fathers for neglecting their upbringing."
>
> Muhammad 'Umar, *Hadir al-Misriyyin*
> *'Aw Sirr Ta'akhurihum* (1902)[49]

In the midst of his long diatribe attacking the Egyptian poor in his 1902 book, Muhammad 'Umar spends a few pages describing street children. As in the rest of his book, he associates public loudness, screaming, and swearing almost exclusively with the "lower" class. This direct connotation of loudness and public "noise" as characteristic of the Egyptian masses was a common stereotype almost universally invoked in most newspapers, books, and other forms of cultural production. 'Umar, however, couldn't bring himself to blame street children for their loudness, vagrancy, and destitution. Instead, he entirely blames the parents of street children for failing to feed, clothe, educate, and properly raise them. Of course, he failed to realize or admit to himself and his readers that street children, if they were not orphaned or entirely abandoned by their parents, did not belong to families that had the economic means to provide them with such an idealized and fetishized vision of "proper" public comportment.

Fortunately, other commentators during the first few decades of the twentieth century were somewhat more realistic and had more compassion when commenting on street children. Unlike declaring common beggars to be imposters, it was much harder to vilify orphans and other street children and much easier for them to be classified as deserving poor. Because of their young age, they were also seen to have a chance of redemption through education or through vocational training. Even the dreaded and much maligned 1923 vagrancy law declared in Article 31 that the "law does not apply to women or children under 15 years of age." Also, more charity associations and fundraising efforts existed to house and feed street children. As Mine Ener has shown, this was in part because the "rhetoric of saving children also served to affirm the paternalist role of the Egyptian upper classes: in the absence of 'proper' care from the children's own parents, Egypt's elite assumed tutelage of children."[50] This did not mean that street children were not attacked or scapegoated in the press—as articles on thefts by street children abounded—but often the blame was assigned to their parents, if they had parents, or to the state that had failed them. For those children suspected of committing crimes, there were juvenile reformatories, and in 1908, legislation was implemented to forcibly confine some street children.[51]

When a group of five homeless young boys were caught stealing hubcaps from cars parked in public areas, it made for a sensationalist story in *al-Ahram*. The paper made sure to photograph the five boys together next to an automobile, and ran the article on the morning of January 14, 1934, under the exaggerated headline: "A Gang of Young Boys Specializes in Stealing Car Accessories" (see Figure 2.5).[52] The five boys were arrested for the theft of hubcaps from cars parked in front of movie theaters and on other public roads. All of the boys were from the Sayyida Zaynab district, and according to the newspaper report, they were all either orphans or abandoned by their families. To the chagrin and surprise of the reporter, one of the boys, the apparent ringleader, had graduated from primary school. It was in part because of their economic desperation and bad social situation that they had gotten together and formed their own street family. Tracking the boy's routine revealed that most of Cairo was their playground, as they walked and most likely tram-surfed a large number of its streets every day.

The article detailed how they usually slept on the sidewalks of some of the smaller side streets near the Sayyida Zaynab Mosque. For an early breakfast, they typically walked westward for a mile or so to Qasr al-'Ayni Hospital, where they collected some of the leftover food given away or thrown out from the day before. On a typical day, they then walked or tram-surfed northward toward the modern and wealthier parts of the city, looking for the desired car models. In the afternoon, they usually stopped at the Rud al-Farag vegetable market (three and a half miles north of the hospital), where apparently some

FIGURE 2.5. A Gang of Young Boys.
SOURCE: *Al-Ahram*, January 14, 1934.

of the more generous food merchants gave them some of the leftover food. It was on their return journey to Sayyida Zaynab in the evenings that they hit their prime spots for more expensive automobiles, those parked between Midan Sulayman Pasha (Tal'at Harb today) and Bab al-luq. There, they could pick off the hubcaps of dozens of cars that were regularly parked in this busy movie theater and shopping district of Cairo. They then sold the hubcaps, for a fraction of their value, at a local gas station back in their home district of Sayyida Zaynab. The photograph suggests that the boys ranged from ten to fifteen years old, but the article indicated that they were being examined by a court doctor to determine their exact age. This assessment was most likely done to officially determine the duration of their forced incarceration in one of Cairo's juvenile reformatories.

Most orphanages and juvenile reformatories provided educational or vocational training, which, in theory at least, was supposed to provide resident children with useful skills to help them acquire future employment. Aside from these programs, there were some experimental projects that the Egyptian government and private donors planned in order to "help" street children. Though to be sure, many of these were not implemented, or if they were, they were deeply flawed. In the spring of 1938, one of those dubious experimental programs was initiated by the Mayor of Cairo, with the encouragement and partial financial support of King Faruq. A detailed editorial in *al-Musawwar* magazine reported on this program and interviewed the Mayor, its sponsor and primary advocate. The editorial and interview, which perfectly captured some of the elitist patronizing attitudes of the time, begins by declaring that "the esteemed 'Abd al-Salam al-Shadhli Pasha canceled the summer vacation program for street children. Instead, he will save these vagrant children by gathering them from the streets of the capital and sending them to be raised in agricultural camps built especially for them."[53] In the interview, al-Shadhli elaborates on how, as Mayor, he set up a special police unit in order to collect Cairo's street children and provisionally placed them in collection centers and orphanages in the capital. When the farming camp was partially built, in the province of Daqahliyya, two hundred children were sent there. The other four hundred children were temporarily kept in orphanages in Cairo until the construction of the camp was complete.

In the interview, the Mayor declared, "we thought about the future of these homeless children and started planning and building this camp. By employing the able among these children and teaching them how to farm we can guarantee that they will not return to the city streets." Although teaching children how to farm may appear to be a worthy cause, the way that these children were forcibly "collected" by the police and eventually placed in these far-off camps should have raised some serious questions about this entire enterprise. The article never addressed what sorts of crops were grown in these camps, or whether these crops were intended to be sold or not. These questions would have opened up a slew of further questions about what happened to any proceeds from the project. The reporter, however, did ask the Mayor about the other summer resort camps that were canceled and replaced by these agricultural camps. Without flinching, al-Shadhli smugly responded: "Would any of those so-called resorts be better than these agricultural camps? To start with, the agricultural camps are located in a completely healthy environment and they are very comfortable for the children and I am sure that this will make them very happy. With this initiative, we are also hitting two birds with one stone, for we are saving the kids from the streets and at the same time they will be enjoying their summers and their winters as well, in complete happiness." The way that al-Shadhli adamantly defends this project reveals his awareness of how suspect the idea of putting children to work in agricultural camps was. Though it would seem that silencing and permanently moving vagrant children far away from the streets of the capital overrode any ethical concerns he might have had on the questionable methods employed in achieving that goal.

Removing vagrant children out of the urban streets was not just a Cairene policy. In the same issue of *al-Musawwar*, yet another article addressed homelessness and street children, under the title "Alexandria follows the lead of Cairo and its fight against homelessness and caring for street children."[54] The short article described how, like the police in Cairo, Alexandria's police had also redoubled their efforts to detain many street children and place them in orphanages. Framing the article are three before and after photographs depicting some of the street children whom the Alexandria police detained. The first picture displayed the children as they were "found": dirty and with torn clothes. The second and third pictures showed them in clean white clothes, though visibly and conspicuously guarded by several police officers. The article concluded that "if the other provinces followed the example of Cairo and Alexandria and expanded their police efforts toward this goal, then we would cleanse the country from all vagrants by providing for them, and making life easier for all of these unfortunate souls." On the surface, it would seem that both the *al-Musawwar* coverage and the public policy initiatives concerning street children and the itinerant poor were struggling to justify these policies as purely acts of institutional benevolence. Charity aside, "cleansing" the country of the itinerant poor (including children) by silencing and removing them from public view also served other imperatives, the most important of which were (ultimately futile) attempts to provide "order" and to create quieter, more sterile, and hence more "civilized" public streets.

Conclusion

> The bourgeoisie and the capitalist system thus experience great difficulty in mastering what is at once their product and the tool of their mastery, namely space. They find themselves unable to reduce practice (the practico-sensory realm, the body, social-spatial practice) to their abstract space, and hence new, spatial, contradictions arise and make themselves felt.
>
> Henri Lefebvre, *The Production of Space*[55]

In many ways, Henri Lefebvre's assessment of the difficulties the technocratic bourgeoisie had in understanding the embodied intricacies of public space speaks directly to the realities of the Egyptian streets. An embodied, more sensory-aware examination of space nuances and tempers any exaggerated bird's-eye view of the panoptic power of the modern state. The plans, maps, laws, and controlling tools and techniques of modern authority may have looked impressive and all-encompassing on two-dimensional sheets of paper, but the actual enforcement, if it ever took place, was almost always contentious. The varying and creative ways in which ordinary Egyptian men and women clashed with or evaded the authorities, through reappropriating the streets for their own use, are the best demonstration of the flaws of exclusively relying on the abstract "theoretical (epistemological) realm" at the expense of the everyday "space of people who deal with material things."[56] To be sure, this does not diminish the very real imbalance and abuse of power often enforced by the police and state authorities, but it reveals some of the cracks and weaknesses of these modern systems of control, especially when it comes to controlling the streets. Listening in to the sources, and closely examining the streets at a more embodied microlevel, gets us closer to what Lefebvre labels the "practico-sensory realm."

As this chapter demonstrates, the itinerant poor and especially street hawkers, individually and at times collectively, defied the state authorities and in many ways appropriated the streets for their own use. In part because of the transitory nature of the itinerant poor, their relationship with the Egyptian state was (and still is) often adversarial and rife with tension. As we have seen, there were official crackdowns, and—for lack of a better word—unofficial shakedowns of street merchants and the itinerant poor by some members of the police. Though it was impossible to consistently regulate, as a totality, tens of thousands of constantly moving hawkers, entertainers, and beggars who, by definition, did not have a known street address. Nonetheless, these cat-and-mouse games between the Egyptian authorities and the itinerant poor were not just about ownership and control of the public street, they were also critical in driving discourses of class formation and defining the shifting sensory sensibilities of upwardly mobile, middle-class Egyptians. Increasingly, as the twentieth century roared on, the Egyptian press was full of classist condemnation and sensory marginalization of the itinerant poor. Silencing the streets and street people and rendering them invisible were critical components of chasing after an elitist ideal (albeit an unrealistic one) of what modern and "civilized" society was supposed to look and sound like.

Infrastructure, Technology, and the Sounds of Modernity

Electric tram-cars now rush boisterously through the streets of Cairo, filled with people who never understood the "go fever" until the advent of the street-railway, two or three years ago. . . . In every direction—to Bulak, the citadel, Abbassieh, through the Ismaileh quarter, even to the site of ancient Fostat—the cars run. . . .
The people have learned the intricacies of "transfers" and "round trips," and their satisfaction over the street traction enterprise, doing more than all other agencies to obliterate the Cairo of old, seems sublime.

        Frederic Penfield, *Present-Day Egypt* (1903)[1]

If you are lucky enough to attend the opera, or go to an evening cinema or a casino/dance hall, do not forget that hundreds of thousands of Cairenes work long days and in the evenings [and nights], they have earned a long overdue rest so they can resume work [the next day]. Do not disturb them with your use of the car horn as you hurriedly drive back to your residence.

        Nizam al-Murur bil-Qahirah [Egyptian
        Traffic Police bulletin] (1938)[2]

3 Roads and Tracks
Modern Traffic and the Sensory and
Social Impact of Trams and Automobiles

For most of the world, the sounds of modernity came roaring through at the turn of the twentieth century, becoming "integral to the phenomenon of modernity itself."[1] These sounds, emanating from a host of new electrical and gasoline- and steam-powered machines, eventually and unevenly transformed first the soundscapes of the streets in urban Egypt and then the sounds of the Egyptian countryside as well. Wherever train tracks were laid, the sounds of train whistles and loud steam engines added to the din. Water pumps, and later on buses, trucks, automobiles, and eventually electricity, would gradually reach small towns and villages in the delta and all along the Nile valley. The soundscapes of Egyptian towns and cities loudly changed as urban morphologies were transformed from the mid-nineteenth to the mid-twentieth century through mass urbanization, modernization, and dramatic changes in urban infrastructure and transportation networks. The acoustics of wider streets and large squares differed from those of the small alleys of the older more traditional urban areas, as broader streets allowed for mass congregations of commuters, pedestrians, and merchants. As the streets were converted from packed earth to paved cobblestone and later on asphalt, the sounds of animal and wheeled traffic also changed.[2] Newer and wider streets and boulevards allowed for wheeled vehicles, from mule- and horse-drawn carts and carriages in the nineteenth century to motorcycles and automobiles in the twentieth century and beyond. Starting in the late 1890s, electric tramways became fully operational, and came to dominate public transportation in Cairo, Alexandria, and Port Said for almost half a century. The internal combustion engine and the spread of automobiles, buses, and motorcycles with their roaring engines, honking horns, and gasoline aromas sensorially added to the new cacophonies and smells of the city, eventually overtaking tramways by the second half of the twentieth century.

This chapter begins by documenting the planning and building of some of these infrastructural changes, while also broadly surveying the impact this may have had on a wide spectrum of Egyptian society. The widening and expansion of roads and the creation of new roads, tracks, and modern infrastructure was uneven, lacked a centralized vision, and was very costly. Next, I survey the types and numbers of new vehicles using this rapidly developing infrastructure, from horse-drawn carriages and motor vehicles to the growing networks of urban electric tramways. Tramways in particular played an important role in centralizing Cairo and Alexandria into more cohesive cities, though simultaneously, they also expanded the peripheries of both cities into newer expansive and sprawling suburbs.

The second half of the chapter examines the changing sounds of city traffic and the resulting hubbub in these new urban spaces. Here and in the remainder of this book, my focus will continue to be on the transformation of the Egyptian streets, squares, and other public spaces and, even more importantly, on the varying ways that ordinary Egyptians benefited from, adapted to, or when necessary resisted those changes. I end this chapter with a more intimate examination of tramway cars as moving, liminal spaces—neither private nor entirely public—where embodied socioeconomic interactions daily took place. In many ways, this chapter and the next—which examines the impact of electricity, street lighting, and Egypt's burgeoning nightlife—serve as this book's backbone, as they both cover a host of important technological and infrastructural changes that set the tone and direction of the rapid changes in Egypt's urban environment, both sonically and in many other ways. From the late-nineteenth to the mid-twentieth centuries, the very materiality of the streets and the physical spaces above and below them changed. Metal tramway tracks and asphalt ran through the streets, allowing for new, often louder, transportation vehicles, pipes carrying water and gas ran underground, and electrical and telephone wires ran above and below, bringing with them transformative sounded technologies unimaginable just decades before. These new infrastructural arteries were (and still are) vitally connected to the social fabric of the city, as they changed how everyday people sensorially, economically, and socially connected with their immediate environment.

Wider Paved Streets and the Introduction of Wheeled Vehicles, 1840–1940

Think of pavement as a rough sound ribbon whose recording and playback mechanisms are shoes, hooves, wheels, canes, pelting rain, hail, and other falling objects. For each groove, rut,

> or pothole worn into the surface, the pavement returns a
> vibration peculiar to the shape and depth of the fissure and to
> the weight, pressure, pace, and contact-contour of each footfall,
> hoof-beat, or wheel-spin. In this sense, pavement is more
> sonically dynamic and temporally responsive than any
> gramophone.
>
> <div align="right">Hillel Schwartz, <i>Making Noise</i>[3]</div>

In the eighteenth century, horse- and mule-drawn coaches and carts were almost nonexistent in Egypt.[4] It was not until wider roads were built in the mid-nineteenth century that wheeled vehicles began to proliferate in the urban centers.[5] Starting in the mid-nineteenth century, the introduction of wider and straighter roads allowed wheeled, animal-powered vehicles in some of Egypt's urban areas.[6] Because most of nineteenth-century Alexandria was newly built, it was ahead of Cairo when it came to having wider streets and modern urban infrastructure like gas street lighting and municipal water.[7] The Suez Canal cities, especially Port Said and Ismailia, were mostly a product of the last third of the nineteenth century, and their street layouts were already spacious enough to accommodate vehicular traffic.[8]

The widening and paving of streets in the long nineteenth century was not done solely for visual, aesthetic, or even practical traffic reasons. Government officials also cited improvements to public hygiene as an important justification for these measures. As Khaled Fahmy has shown in his analysis of the modernization of nineteenth-century Cairo, many of the "strong measures eventually taken to eradicate sources of foul smell considerably affected the shape of the city."[9] Fear of epidemics, which sometimes turned into an obsession, was an important impetus for widening the streets and modernizing the city. In many of these cases, the olfactory sense was invoked and mobilized to motivate government action to widen and clear Cairo's streets.[10] The Egyptian and colonial authorities, aided by endless articles in the Egyptian press, continued to use the maintenance of public hygiene and public order as the primary excuse for cracking down on the itinerant poor. These campaigns to clear and clean up the urban streets often invoked a multisensory offensive on beggars and many street hawkers, who were deemed a danger to the health and the peace and tranquility of the public.

During the reign of Khedive Ismail (r. 1863–1878) and mostly under the direction of Ali Pasha Mubarak, three modern districts were built to the north and west of Cairo, extending the original city toward the Nile. The Ismailia, Azbakiyya, and Nasiriyya districts, which today make up Cairo's downtown area, had wide streets and large squares (*mayadin*, sg. *midan*), and soon would be equipped with municipal water and gaslight. Electricity and the electric tram would also be introduced just a couple of decades later, and multiple new bridges would cross the Nile, expanding the city through roads, trains, and tramway lines further west past Gezira (Zamalek) and toward Giza and Imbaba.

The Azbakiyya district was developed first, with Azbakiyya Garden at its center and three major city squares near three of the garden's four corners[11] (see Map 3.1). Al-Ataba al-Khadra Square, located near the southeast corner of Azbakiyya Garden, would soon became the most vital transportation nexus in Cairo, especially after the Cairo tramway company strategically placed its main tramway terminus in the heart of the square.[12] Muski Street began from the northeast corner of Ataba Square and traveled almost directly eastward, eventually turning into what was then called al-Tariq al-Gadid (the New Road, or Rue Neuve), which was painstakingly and destructively built through old Cairo until it reached past al-Husayn district, which included the Khan al-Khalili bazaar and al-Azhar University. Muhammad Ali Street also connected from the southeastern corner of Ataba Square, extending at a forty-five-degree angle toward the southeast, until reaching the Sultan Hassan Mosque and the Cairo Citadel.

The newly built parts of Cairo already had wider streets, more than sufficient for wheeled traffic, but widening preexisting streets in the older, more populous parts of the city proved very difficult, as the state had to resort to expropriating land and demolishing many preexisting buildings and storefronts. This caused delays owing to legal battles and negotiations with owners for legal compensation. For example, Muhammad Ali Street—which was started in 1845 though not finished until a couple of decades later—was an important attempt at creating better circulation in order to allow regular vehicular traffic into the heart of the old city. It was a difficult and politically sensitive process requiring the forced acquisition and demolition of hundreds of buildings along the way, including "baths, shops, and mosques."[13] This was also the case for al-Tariq al-Gadid, which was carved through a highly populated part of old Cairo.

In the first few decades of the twentieth century, road planning and building was not very centralized. The Tanzim Councils and the Public Works Department, of course, had a great deal of influence, but there were also a host of other independent municipal and government entities that had a say in the planning

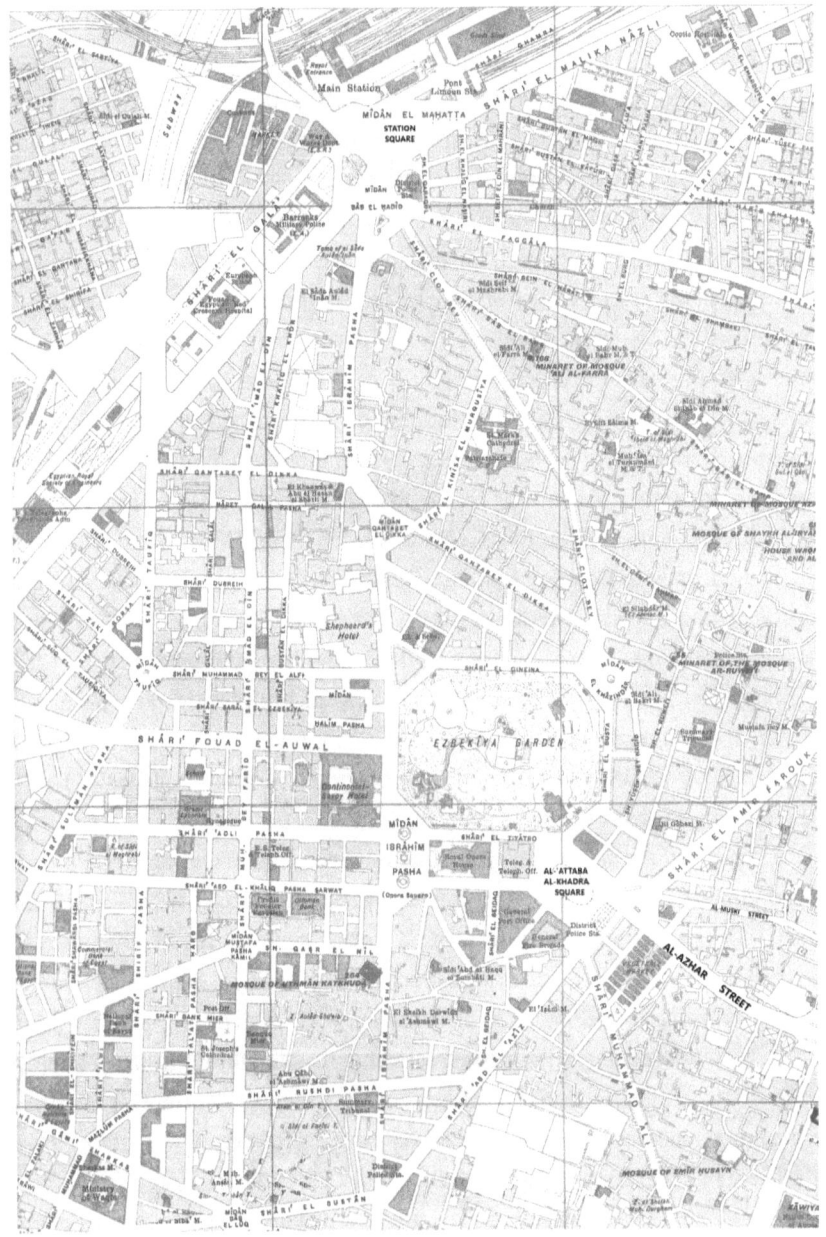

MAP 3.1. Cairo's Roads and Squares.
SOURCE: Adapted from *Survey of Egypt: Map of Cairo* (Cairo: Maslahit al-Misaha [Survey Department], 1950).

and building of roads and other infrastructural projects.[14] In late-nineteenth-century Cairo, for instance, the Water Company conducted some of the street planning. The Roads Department (Maslahit al-Turuq), which was under the jurisdiction of the Ministry of the Interior, also had a say; so did the Department of Public Parks.[15] Many private companies, businesses, and especially utility and transportation companies also played a role in the planning and building of roads. Tramway companies, for instance, were obligated to pave and light the streets adjacent to their tramway tracks.[16] Sometimes, wealthy merchants paved roads in the immediate vicinity of their businesses. In late 1860s Alexandria, as On Barak has shown, the city's wealthy export merchants, who stood to benefit from better roads that would allow them to quickly and efficiently transport cotton to the ports, took matters into their own hands and built their own granite-paver roads in the immediate vicinity of the Alexandria port. In the calculation of Alexandria's Commission of Export Merchants, spending up-front money to improve road infrastructure, increased the speed and the traffic of wheeled carts, streamlining the merchants' business operations and saving them money in the long term.[17] Although this solution may have worked well for the merchants involved, it illustrates the hodgepodge approach and the diversified and independent agendas of the many private, business, and public actors and institutions involved in the planning and building of roads and infrastructure in Egypt.

Despite such structural and planning limitations, urban and rural roads were rapidly built across Egypt. The area covered by roads in Cairo alone, doubled from 2,303,482 square meters in 1902 to 4,646,665 square meters in 1921.[18] Also, whereas it was difficult to travel by road in a wheeled vehicle from Cairo to Alexandria in the late nineteenth century,[19] by 1912, more than 8,500 LE had been spent on widening, leveling, and adding to the existing agricultural road, allowing a direct roadway connection between the cities. At that time, most of the road, like most of the other agricultural roads in Egypt, was made of compacted earth, except for short stretches that were macadamized and "paved" close to the cities.[20] By the late 1920s, it was possible to travel from Alexandria to Luxor by automobile, though not very comfortably, as the road was unpaved for most of the way. In the 1920s, the best roads for motor traffic were in Cairo, Alexandria, the canal cities, and their immediate surroundings, though for the next couple of decades, better roads and highways were gradually built beyond those cities.[21] Building these roads, however, was only a first step, since maintaining them and changing their surfaces to accommodate increasing traffic proved even more challenging. Keeping up with the rapid rise in the number of trucks, buses, and automobiles was an engineering, policing, and infrastructural challenge that stretched the budgets of most cities in the 1920s.[22]

Road building in Egypt greatly increased after the signing of the 1936 Anglo-Egyptian Treaty. The treaty extended Egypt's nominal independence, and the country would soon join the League of Nations and abolish the nineteenth-century capitulations. The British also agreed to withdraw their troops to their bases near the Suez Canal within the next few years.[23] One of the key clauses of the treaty addressed the building of "treaty roads," to facilitate British troop movement from the Suez Canal bases to help defend Egypt and the canal in case of a foreign attack.[24] The British increasingly viewed the Italian presence in Libya, as a direct threat to Egypt and the Suez Canal, especially after the 1935 Italian invasion of Ethiopia and Italy's rapprochement with Nazi Germany. The building of these highways and roads was a high priority, and construction began in earnest. It was of course a costly project, since the roads had to meet rigid military specifications in order to support heavy tanks, trucks, and artillery and personnel carriers. In 1937, the Egyptian Ministry of Communications estimated that "the approximate cost of all Treaty roads can be put at £E 2,000,000 for road-making, and at £E 750,000 for road bridges."[25] The coastal railway from Alexandria westward was extended to Marsa Matruh as early as February 1936, and work on building a paved coastal road to Matruh and eventually Sallum on the Libyan border was accelerated. By late 1937, construction had already begun on an Ismailia to Alexandria road, cutting through the delta; an Ismailia to Cairo road, a Port Said to Suez road, and an Alexandria to Cairo desert highway.[26] This was just the beginning, as the impending arrival of World War II necessitated the building of more roads and bridges throughout Egypt. Many of these roads, of course, were used by civilian traffic well after the war, providing the basis of a basic highway system connecting most of the northern cities via motor road.

Tramways and the Integration of Urban Areas, 1896–1950

> In the street, omnibuses and tram-cars rumble by, blowing strident horns; but the passengers who sit on the seats beneath the awning are not Europeans they are Egyptians, effendis, clerks, shopkeepers, sheykhs [shaykhs], often simple fellahin come to town on business and driving in from Bulak or Kasr-en-Nil.
>
> Stanley Lane-Poole, *The Story of Cairo* (1902)[27]

Soon after electricity was introduced to Egypt, electric tramways would arrive on the scene, drastically transforming Cairo, Alexandria, and Port Said.[28] Electric tramways created an affordable and reliable means of local urban transportation. In Cairo at the turn of the century, first-class tickets cost only six millimes (three-fifths of a piaster) and second-class tickets cost four millimes (two-fifths of a piaster), making tramways affordable to most Egyptians.[29] This allowed a growing number of urban workers to commute to their places of employment in and out of the cities. A daily commute connecting workers to factories, workshops, offices, and shops helped to establish regular traffic patterns, setting a sort of audible and routine internal clock for Egypt's growing cities.[30] Also, in many ways, the tramways started a process of urban expansion and suburbanization, especially in Cairo and Alexandria. Alexandria expanded eastward from Raml (Ramleh) Station toward Stanley, Sidi Bishr, and beyond, and Cairo expanded northward toward Shobra, Abbasiyya, and eventually Heliopolis, southward toward Maadi, and westward across the Nile to Zamalek, Giza, and Imbaba.[31]

The concession for a Cairo tramway company was signed on December 5, 1894, after negotiations between the Egyptian government and the Belgian Société Générale des Chemins de Fer Économiques, officially forming the Société Anonyme des Tramways du Caire in the spring of 1895. Some of the key specifications for this first tramway concession were (1) a maximum speed limit of ten kilometers per hour, (2) at least forty daily trips per route, (3) no services between the hours of 1:00 a.m. and dawn, and (4) the designation of a "women only" compartment in each tram.[32] It is important to note here that the women only compartment was just for first-class customers. There were no "men only" compartments, so working-class women rode along with men in the more affordable second-class sections. Therefore, during the first couple of decades of trams in Egypt, there were in effect three compartments: (1) a women only first-class section, (2) a mixed sex first-class section, and (3) a mixed sex second-class section for the rest of the population. As I examine in later chapters, for Egypt's upper and upper-middle classes, until the gradual gender desegregation starting in the 1920s, mixing with the loud, "unwashed masses" was as threatening as gender mixing.[33]

The first electric tramway was inaugurated in an official ceremony in Cairo on August 12, 1896, with Egyptian, Belgium, and other foreign dignitaries ceremonially taking the well-advertised first trip.[34] This first line, which soon opened to the general public, ran from the bottom of the Cairo citadel and near the Sultan Hassan Mosque, cutting through Muhammad Ali Street to what would become the main tramway junction at al-Ataba al-Khadra Square, then heading due west to Bulaq near the Nile. New lines would soon open, reaching out to the north and west from Ataba Square. The first Abbasiyya line opened up in September 1896 and was extended in January 1898.[35] Other lines would also soon open up from Ataba Square to the Cairo Rail Station, Bulaq, Qasr al-Nil, Sayyida Zaynab, and elsewhere. On May 20, 1903, two new lines connected the tram network northward toward Shobra and Rud al-Farag, and the maximum speed limit for this line was increased to twelve kilometers an hour, which would soon become the speed limit for the rest of the tracks in urban Cairo.[36]

In the summer of 1897, the first tramway on the west coast of the Nile was extended for 12.4 km, from Giza, near the Nile, to the pyramids, ending right by the Mena House Hotel at the foot of the Pyramids Plateau. Because this line ran through the countryside, the maximum speed limit for the Giza tram was set at thirty kilometers per hour. This line was not connected to the main city lines on the east bank until January 2, 1909, when it was linked through the Abbas Bridge and Roda Island. Another tramway Nile crossing was added in 1912 from Abu al-'Ila Bridge, connecting Bulaq and Imbaba through the island of Zamalek.[37] These were the early beginnings of the growth of a greater Cairo, incorporating the Nile islands and the west bank of the Nile into the city and the east bank.

The expansion of Cairo in a northeasterly direction would also take place primarily through a tramway line. In a bold investment, the Belgian tramway tycoon Baron Édouard Louis Joseph Empain, one of the directors and part owner of the Société Anonyme des Tramways du Caire, planned and partly funded a completely new suburb of Cairo. Heliopolis, designed to be a playground for Cairo's elite with all the modern amenities and recreational facilities, was located in the middle of the desert, about ten kilometers from Cairo. In 1905, Empain created the Cairo Electric Railways and Heliopolis Oases Company, which was officially a subsidiary of Tramways du Caire, and the building of Heliopolis and its fast-moving metro tramway line began in earnest.[38] Heliopolis' population would reach 16,000 inhabitants in 1914, and it would continue to expand at a steady pace in the next couple of decades.[39]

In the post–World War I period, Cairo's tramways continued to improve and attract more customers and revenues (see Table 3.1). In 1923, two major roads started to be built from Ataba Square through the old city, at the hefty cost of 800,000 LE. The fifty-meter-wide Prince Faruq Road connected Ataba Square to Abbasiyya in the northeast, and the twenty-meter-wide Azhar Street paralleled Muski Street eastward toward al-Azhar University. These two new roads provided a perfect opportunity for the only major expansion of Cairo tramway lines in the interwar period. The Prince Faruq Road tramway line opened on April 10, 1929, and the Azhar Street tramway line opened a year later, on June 12, 1930.[40] More than the growing network of wider roads described earlier, the trams helped to integrate the urban areas of Cairo, Alexandria, and Port Said, fostering the formation of more coherent and centralized cities. The Cairene tramway system transported almost 10 million passengers yearly in 1899, over 51 million passengers in 1909, and 82 million passengers in 1920, and reached over 178 million passengers in 1950.[41] The rates of increase in the number of passengers were similar in Alexandria as well (see Table 3.1). Trams were reliable, regular, and affordable for most Egyptians and foreign residents alike, and by mid-century, they became the transportation of choice for the vast majority of Cairenes and Alexandrians. Yet tramways were not simply a means of transportation; commuting and riding the tram provided an entirely new social experience for most of the passengers. As I discuss later, tramways and tramway stations spatially and sonically added a different dimension to the cultural, social, and economic fabric of urban areas.

Carriages, Cars, and Tramways: City Traffic and the Changing Sounds of the Road

Carriages and Carts

In large part because streets were not wide enough to accommodate them, during the first third of the nineteenth century, wheeled vehicles were largely absent from Egyptian towns and cities.[42] As I examined earlier, this began to change at mid-century, and horse-drawn coaches and mule- and horse-drawn carts became a common sight in Cairo and Alexandria. The newer canal cities, of course, already had wider roads, allowing larger wheeled vehicles from their inception in the mid- to late nineteenth century. At first, horse-drawn cabs were used by wealthy Egyptians and foreigners, though they become more affordable to the middle classes as their numbers increased by the end of the nineteenth century. In the meanwhile, donkey- and mule-drawn carts, which were also used to transport goods, would

TABLE 3.1. Electric Tramways in Cairo and Alexandria (1899–1950).

	YEARLY PASSENGERS	KILOMETERS PER YEAR	GROSS INCOME (in Egyptian Pounds, LE)
Cairo Tramway			
1899	9,856,699	1,902,234	—
1904	27,974,000	6,234,000	143,295
1909	51,736,000	11,207,000	270,151
1914–1915	58,224,000	11,013,000	263,215
1915–1916	50,489,000	9,894,000	292,128
1918–1919	74,767,000	10,125,000	369,478
1919–1920	57,616,000	7,353,000	379,781
1920–1921	82,733,000	9,844,000	386,663
1921	69,808,000	8,967,000	—
1935	88,957,000	17,479,000	553,796
1950	178,650,000	15,938,000	1,461,532
Alexandria Tramway			
1904	11,672,000	2,179,000	61,093
1909	16,637,000	2,993,000	88,123
1914–1915	17,104,000	5,049,000	85,799
1915–1916	18,186,000	5,687,000	120,519
1918–1919	32,791,000	7,870,000	158,715
1919–1920	29,861,000	3,693,000	197,604
1920–1921	31,621,000	4,946,000	227,018
1935	29,861,000	8,378,000	155,578
1950	87,146,000	6,442,000	569,359
Alexandria and Ramleh Railway Co.			
1904	4,307,000	1,050,000	49,884
1909	8,706,000	2,544,000	71,487
1914–1915	11,278,000	2,911,000	83,333
1915–1916	14,730,000	4,668,000	111,695
1918–1919	19,591,000	4,854,000	119,654
1919–1920	17,403,000	2,078,000	145,083
1920–1921	20,087,000	3,339,000	186,662
1935	23,453,000	6,814,000	131,158
1950	44,303,000	3,888,000	518,735

SOURCE: *Annuaire statistique de l'Egypte 1920–1956*, Ministère des finances, Direction de la statistique (Cairo: Imprimerie Nationale, 1957).

become the ride of choice of lower- and lower-middle class Egyptian men and women. Later in the century, mule- and horse-drawn omnibuses were introduced in Cairo, Alexandria, and the canal cities, and they became popular with most Egyptians who could not afford coaches and other means of transportation. By 1932, Egyptian roads supported at least 50,000 of these animal-powered carriages. Over two thirds of these vehicles were used in Cairo and Alexandria alone (see Table 3.2). Moreover, these numbers reflect only the vehicles registered and licensed with the government. The actual figures were likely much larger.

It is not hard to imagine the sonic implications of these dramatic shifts in just transportation technology alone. The introduction and proliferation of wheeled animal-drawn vehicles undoubtedly changed the sounds of the streets. The sounds emanating from simple dirt roads were muffled, macadam roads were louder, and cobblestone streets were louder still, producing a unique clacking and rumbling sound.[43] As asphalt-paved roads became more common in the 1920s, the sounds from carriages and automobiles also changed.[44] Wooden wheeled carts and coaches produced unique creaking sounds as they rolled by in the street. These sounds, however, would change markedly depending on the speed of these vehicles and on the type of road surface they traversed.[45] The sounds of the galloping or trotting horses or mules added to the din, along with the occasional cracks of drivers' whips. In the early days of private coaches, a *sayis* (usually a boy or a young man) would walk or run ahead of the coach to warn and move pedestrians and other traffic out of the way. According to John Chalcraft, employing a *sayis* was still a standard practice in the 1870s.[46] Also, most of the horses, mules, and donkeys pulling these vehicles wore small bells jingling loudly, mainly to alert pedestrians and other vehicles to their presence. Loud whip cracking also served to warn pedestrians and to signal the horse to go faster, and as I discussed in Chapter 1, when necessary, the cab or cart driver would yell out to warn others of his impending approach. Nonverbal vocalizations by carriage drivers were also common. Loud whistling, hissing, and tongue and cheek clicking was often used to signal to the animal and to pedestrian traffic. In some cases, cart drivers beat a stick against their cart, producing loud drumming sounds to alert pedestrians.[47] The growing number of carts and carriages would soon compete for road space with trams and bicycles in the late nineteenth century and motor vehicles at the turn of the twentieth, all of which significantly added to the sounds and hubbub of the streets.

TABLE 3.2. Licensed Animal-Drawn Vehicles in Egypt (1891–1953).

	CAIRO	ALEXANDRIA	TOTAL IN EGYPT
Horse-Drawn Carriages (Cabs)			
1891	—	1,516	—
1916	1,586	2,199	—
1921	1,321	1,666	—
1932	732	1,320	3,332
1936	1,009	1,143	3,826
1947	698	943	3,322
1953	386	747	3,244
Animal-Drawn Flatbed Carts/Wagons (Karro)			
1891	—	1,935	—
1906	—	3,357	—
1916	11,289	3,038	—
1921	9,000	9,391	—
1932	23,074	13,642	46,566
1936	13,470	10,092	39,222
1947	30,000	4,605	51,619
1953	23,176	5,106	49,056

SOURCE: *Annuaire statistique de l'Egypte 1951–54*, Ministère des finances, Direction de la statistique (Cairo: Imprimerie Nationale, 1956).

Although sidewalks were plentiful in the newly built parts of Cairo, in the more populous traditional areas, sidewalks were at a premium (if they were available at all), and pedestrians had to share the streets with all of the above wheeled vehicles and in some parts of the city they also had to deal with passing trams. This reality necessitated an extensive and constant reliance on intentional and unintentional auditory signaling and notifications—be they in the form of bells, the hooves of galloping horses, whistles, horns, engine sounds, or human voices. To the initiated ear of a local resident, all these sounds were a part of a sonic vocabulary that was innately and subconsciously familiar, allowing easy and intuitive navigation of the streets. However, when new and unfamiliar vehicles were thrown into the mix, it took some adjustment before people's senses recalibrated to the size, speed, "behavior," and sounds of these vehicles. That was certainly the case with trams and motor vehicles, which in the first couple of years of their introduction to traffic streams, presented new challenges to pedestrians and carriage drivers.

The Silence of the (Killer) Trams

> Alexandria seemed a capital city. In the trim garden nurses were rolling their prams and children their hoops. The trams squashed and clicked and rattled.
>
> Lawrence Durrell, *Justine*[48]

Trams are obviously not silent. When their heavy cars moved along the tracks, they "squashed and clicked and rattled," as Lawrence Durrell described in his Alexandria Quartet. There was also a perpetual buzzing sound produced as the metal rod pantograph on top of the tram grazed the overhead electrical wires. However, relative to trains and automobiles with their loud steam and internal combustion engines, the sounds that trams made as they raced in the streets seemed minimal. Also, in a loud and busy urban environment, it could be difficult for many to hear an approaching tram. In one of the frequently quoted anecdotes about early tramways in Cairo, Egyptian children serenaded the first tramway run, in August 1896, by repeatedly chanting, *al-afrit* (the ghost).[49] These children had most certainly seen and heard trains before, as by that time, trains had been operating in Egypt for almost a half century. Therefore, the *al-afrit* chants likely reflected the relative silence of the tram in comparison to a steam locomotive, and not that the "ignorant" native children believed the horseless tram magically propelled itself along the tracks.[50] Nevertheless, seeing and hearing the electric tram for the first time must have been an altogether novel experience. This new experience is somewhat recaptured and comically made light of, by a popular folk song that circulated shortly after the introduction of the electric tram in Egypt:

> The electric car, the electric car . . . It rushed through to Manshiyya
> The electric car is running well . . . But when it stopped in front of Khamis
> Khamis froze up and stared . . . And spilled his okra in his *mulukhiya* [soup]
> The electric car is made of pure brass . . . But when it stopped in front of Elias
> Elias stood there stupefied . . . And spilled his okra in his *qulqas* [taro stew][51]

When trams were first introduced to Alexandria, Cairo, and Port Said, they had frequent collisions with carriages, carts, and pedestrians; some were fatal. Predictably, the relative silence of the trams was blamed as one of the key reasons for some of these inevitable accidents and deaths. Journalists, commentators, and legislators tried to find varying ways of projecting noise to alert pedestrians to oncoming electric trams.[52] Sensationalistic newspaper headlines about the dangers of the tram abounded. Some, no doubt, were nationalistic, influenced by the fact that the foreign-owned tramways were trampling primarily Egyptians.[53] Yet, vilifying the electric tramways for their dangers was, in fact, common around the globe at this time. Because of safety concerns, many Parisians around 1900 apparently called tramways the "murderous tramways" and "tramways of death."[54]

In Egypt, alarmist press coverage of tramway accidents led to some political action. In a letter dated February 7, 1898, Umar Lutfi, the president of the national Majlis Shurah al-Qawanin (the Legislative Council), complained to the president of the Majlis al-Nuzzar (Council of Ministers) that something had to be done about the high rates of tramway accidents in Cairo. The letter begins by announcing that on Thursday, February 3, 1898, the Legislative Council had met to discuss "the grave dangers and painful accidents caused by the tramway company," which according to the council, had led to "the killing of some and the injury and maiming of countless others."

The letter blames the accidents on the lack of safety procedures and the apparent exhaustion of the tramway operators and workers, who according to the council were forced to work "sixteen hours straight, which leads to tired and inattentive workers, causing these accidents." The rest of the letter suggests that the government should stipulate that the tram company install sonic warning devices, have its agents warn people in the streets, and limit the size of the trams to two connected cars.[55] When neither Lutfi nor the Legislative Council received a response to their first letter, the entire council met on April 21, 1898, and wrote up another letter admonishing the president of the Council of Ministers for not responding: "There are many innocents that are killed, children orphaned, and women that are widowed. This is despite the fact that we have previously warned the government about this issue. We urge you to act in order to remedy the situation."[56]

Not much was done when it came to reducing the working hours of tram operators. This would, in part, lead to a long history of strikes and labor activism by tram workers in Egypt.[57] However, because of the uproar against pedestrian deaths, tramway companies in Egypt and elsewhere devoted some time and effort to figuring out ways to sonically warn pedestrians and improve tram safety. Tram horns and bells were employed more regularly, but in the early days of the tram, before pedestrians got used to their pace and sound, accidents were inevitably somewhat frequent. To be sure, decades before, there was a similar outcry from many over trains trampling people and, more often, farm animals.[58] Unlike trains, however, which mostly operated outside the city centers and did not have to deal with city traffic, trams operated in the middle of crowded urban areas and pedestrians needed to adapt to their presence. Also, for the most part, train tracks were typically separated from other traffic, whereas many tramways in Egypt shared the roadways with pedestrians and other wheeled traffic, including the automobiles that would soon be introduced into the cacophonous traffic mix. To be sure, as automobile numbers began to increase during the first two decades of the twentieth century (see Table 3.3), complaints about the dangers of cars and trucks also increased.

TABLE 3.3. Registered Motorized Vehicles in Cairo and Alexandria (1904–1921).
Note: These figures do not include government- or military-owned vehicles.

	CAIRO	ALEXANDRIA
1904		
Total vehicles (1904)	—	18
1911		
Total vehicles (1911)	—	155
1916		
Private vehicles	745	263
Taxis/buses/trucks	192	24
Private motorcycles	457	168
All vehicles (1916)	1,394	551
1921		
Private vehicles	1,944	1,372
Taxis/buses/trucks	497	267
Private motorcycles	930	390
All vehicles (1921)	3,371	2,029

SOURCE: *Annuaire statistique de l'Egypte 1921–1922*, Ministère des finances, Direction de la statistique (Cairo: Imprimerie Nationale, 1923), 77.

Loud and Smelly Motor Vehicles

> I decided recently to ride in a motorized omnibus in order to experience something new. So, I went to a bus stop and rushed to sit in one of the empty seats. As soon as I sat down, the driver began to move a metal rail next to his leg and I immediately heard this awful sound, smelled this disgusting smell, and felt vibrations reverberate in every part of my body. It was as if I, and all the other passengers experienced a sudden illness. However, I remained seated, thinking that I am simply not used to riding this vehicle. Finally, as the machine started moving, billowing smoke, rumbling loudly, while the hated smell of gas began to spread, I prayed for God's help. It was at that moment that the conductor arrived and I bought from him a ticket till the end of the line. . . . Immediately afterwards, the automobile veered towards the left as I fell on top of the man sitting next to me and we both fell on a third passenger and all three of us leaned towards the fourth. If it wasn't for the safety barrier we would have all been goners. . . . We continued in this state of affair for a while. Every time the bus leaned towards the right we all leaned towards the right, when the driver turned left all of us swayed towards the left. It was as if we were all participating in a Sufi *zikr* [ritual prayer]. This continued until we arrived at the Cairo Train Station. As I stepped out of this machine, I proceeded to curse all motor vehicles, their inventors, and their manufacturers.
>
> Fathi 'Azmi (*Al-Dik*, June 17, 1908)[59]

The 1908 newspaper article in which this portrayal appeared was titled "The Worst Invention to Appear in the Twentieth Century," and it specifically describes one of the early "motorized omnibuses" in Cairo, which was probably operated by either the Société des Automobiles et Omnibus du Caire (SAOC) or the Cairo Public Motor Car Service.[60] *Al-Dik* (The rooster) was a weekly satirical periodical, so we must take the words of Fathi 'Azmi as spoken in that context.[61] Yet even though satirical, this multisensory and embodied critique of motor vehicles provides a glimpse of the relative novelty of experiencing an automobile or bus for the first time. It also, even if comically, relates the auditory and sensory overload experienced by many who first encountered automobiles. 'Azmi's flamboyant account, with its descriptions of the foul smells of gasoline exhaust, loud vibrating engine sounds physically reverberating people's bodies, and the inevitable touching of passengers who are crowded in an enclosed space, gives a vivid embodied narrative of what would soon become the everyday reality of urban dwellers everywhere. In 1908, when 'Azmi wrote his satire, automobiles were still rare in Egypt. The entire city of Alexandria, for example, had only 173 motor vehicles.[62] But by the second decade of the twentieth century, the numbers and sounds of automobiles, trucks, buses, and motorcycles began to grow exponentially, creating unexpected traffic and infrastructural problems.

In 1920s downtown Cairo, the obvious though challenging way to deal with this increased traffic was to build newer and wider roads. In 1923, two rather ambitious road projects were started to alleviate the heavy traffic in Muski Street and Ataba Square. What made these two projects difficult and expensive to execute, was that just as with the nineteenth-century construction of Muhammad Ali Street and al-Tariq al-Gadid, the new roads had to pierce through some of the preexisting buildings and alleyways to the north and east of Ataba Square. Expropriating buildings from their owners was always a controversial and expensive proposition. Nonetheless, the political will and the necessary 800,000 LE were creatively cobbled together, and construction started in 1923. Prince Faruq Road (Tariq al-Gaysh today) headed in a northeasterly direction and, according to the *Cairo Tanzim Department Annual Report for 1923–24*, was set to be "50 meters wide, connecting the congested Ataba El Khadra Square with the newer district of Abbassia." The other road, al-Azhar Street, was twenty meters wide and it penetrated eastward, paralleling Muski Street toward al-Azhar University and "providing a relief road to Mouski [Muski] street." Both of these roads were paved with asphalt; however, to save on costs, only the sections assessed to have heavy traffic were built with a concrete foundation, the other sections received an inferior macadam base. In part because of these two roads, the Cairo Tanzim yearly report triumphantly declared that "far greater town improvement schemes have been projected during the last two years than at any previous period since the great Khedive Ismail."[63] Hyperbole aside, the Prince Faruq Road was vital, as it had the added benefit of relieving traffic on the always-congested Station Square (Bab al-Hadid Square; known as Ramses Square today), where Cairo Station (Mahatit Misr), the main railway junction for Cairo, is located. At the time, all the traffic to Abbasiyya, Heliopolis, and beyond, had to pass through

Station Square. According to the Tanzim report, "The completion of the new Abbassia[Abbasiyya]–Ataba route will greatly alleviate traffic congestion not only in Station Square, but in Clot Bey & Nubar Pasha streets, and will further reduce the distance between Abbassia and Ataba by 50%."[64]

Building new roads, however, was not enough to alleviate many of the unexpected problems that the ever-increasing numbers of cars and especially trucks and buses produced. As the twentieth century roared on, the mounting number of motor vehicles not only proved destructive to the still mostly macadamized roads, but they also put an unexpected stress on all the preexisting road, traffic, and policing institutions. As early as 1920, police officials in Cairo and Alexandria were complaining about increased motor vehicle traffic. Describing Cairo, the *Ministry of the Interior Report for the Year 1920* declared: "The difficulty of adequately controlling the traffic in the city is largely increased by the growing number of motor-cars on the streets. It will not be easy to effect any great improvement in this respect until heavier penalties can be inflicted on persons contravening the traffic regulations than are possible under the present law." Also in Alexandria, according to the same report, "the number of motor-car licenses during 1920 is nearly double that of 1919 with a consequent increase in the difficulty of controlling traffic in the city."[65] Of course, in hindsight, this apparent panic on the part of some of the authorities over the number of automobiles in 1920 seems rather quaint, as motor vehicles numbers skyrocketed exponentially in the next few decades (see Table 3.4). By the late 1920s, motorized taxis would replace horse-drawn coaches as the new ride of choice for the middle classes, with more of the upper classes buying their own automobiles and hiring their own chauffeurs.

Also in the 1920s, buses began to proliferate in Egypt's urban centers. Schools and hotels employed them first, but soon after they were used for public transportation, competing with tramways and trains for cheap fares for the masses.[66] By 1922, in the "larger provincial towns there were motor-bus services largely patronized by the native population." In Cairo alone, there were seven small bus companies operating in 1923, carrying out "regular services on many routes in direct competition with the trams, and on others through streets not included in the itinerary of the tramway companies." Many of these buses were rigged with "makeshift bodies" that were locally mounted on light truck chassis, which was much cheaper than buying readymade buses. According to a British report, these buses were "proving very popular with all classes and nationalities, and fill

TABLE 3.4. Licensed Motorized Vehicles in Egypt (1925–1954).
Note: These figures do not include government- or military-owned vehicles.

	TOTAL IN EGYPT (APPROX.)
1925	
Private cars	7,661
Taxis	3,739
Trucks	1,246
Buses	848
Motorcycles	3,456
All vehicles (1925)	**16,950**
1936	
Private cars	20,771
Taxis	4,152
Trucks	2,794
Buses	1,103
Motorcycles	1,909
All vehicles (1936)	**30,729**
1946	
Private cars and taxis	30,948
Trucks and busses	9,497
Motorcycles	3,602
All vehicles (1946)	**44,047**
1954	
Private cars	58,808
Taxis	12,221
Trucks	15,156
Buses	5,331
Motorcycles	11,634
All vehicles (1954)	**103,210**

SOURCE: E. Homan Mulock, *Report on the Economic and Financial Situation of Egypt*, Department of Overseas Trade (London: His Majesty's Stationary Office, 1925), 39–40; *Annuaire statistique de l'Egypte 1935-36*, Ministère des finances, Direction de la statistique (Cairo: Imprimerie Nationale, 1938); *Annuaire statistique de l'Egypte 1951-54*, Ministère des finances, Direction de la statistique (Cairo: Imprimerie Nationale, 1956), 265–266.

a long felt want" for cheap transportation.⁶⁷ In fact, as early as 1924, the Egyptian Government Railway had to reduce third-class ticket prices to compete with the number of "frequent, regular and cheap" bus services between Cairo, Alexandria, and their outlying towns.⁶⁸ By 1928, Cairo alone had thirty-three bus routes and "no less than 464 buses actually running in the city" (see Map 3.2). Additionally, there were over 500 buses in the provinces near Cairo, which at the time were operating "within the jurisdiction of the Cairo traffic authorities." These figures do not include the increasing number of private buses owned and operated by "hotels, schools, clubs and religious institutions."⁶⁹

For better or worse, just as in the rest of the world, Egypt's roads and traffic police were not ready or equipped to deal with this rapid and unexpected rise in the number of motor vehicles, and they would continue to play catch-up for years to come. The total number of motor vehicles in Egypt reached around 21,000 in 1927, 31,000 in 1936, and more than tripled to around 103,000 in 1954.⁷⁰ The environmental, cultural, and economic impact of this traffic explosion affected many urban Egyptians on a fundamental level. As I examine in the rest of this chapter, the hubbub of this new motor traffic and its increasingly loud electric horns, along with the noises radiating from trams, animal-powered traffic, and thousands of pedestrians and passengers, created a vibrant yet cacophonous soundscape characterizing all Egyptian cities.

Controlling Motor Vehicle Horn Noises

> Every effort has been made to reduce noise to a minimum. The number of contraventions for excessive use of electric horn or klaxon drawn up during the year show an increase of approximately 1,000, in figures for 1936, while thousands of warning letters have been sent to owners in the cases of first offense.
>
> Annual Report of the Cairo City Police for the Year 1937⁷¹

Tramways, trains, and the steady increase in motor vehicle traffic dramatically transformed the soundscape and decibel levels of Egyptian cities and towns. As elaborated previously, just as with the early introduction of trains and tramways, the Egyptian press printed alarmist articles featuring "killer cars" and "death under the wheels," as pedestrians recalibrated their senses in order to share the road with much faster and heavier vehicles. The lack of sidewalks in many of the poorer urban areas also exacerbated this problem. For this reason,

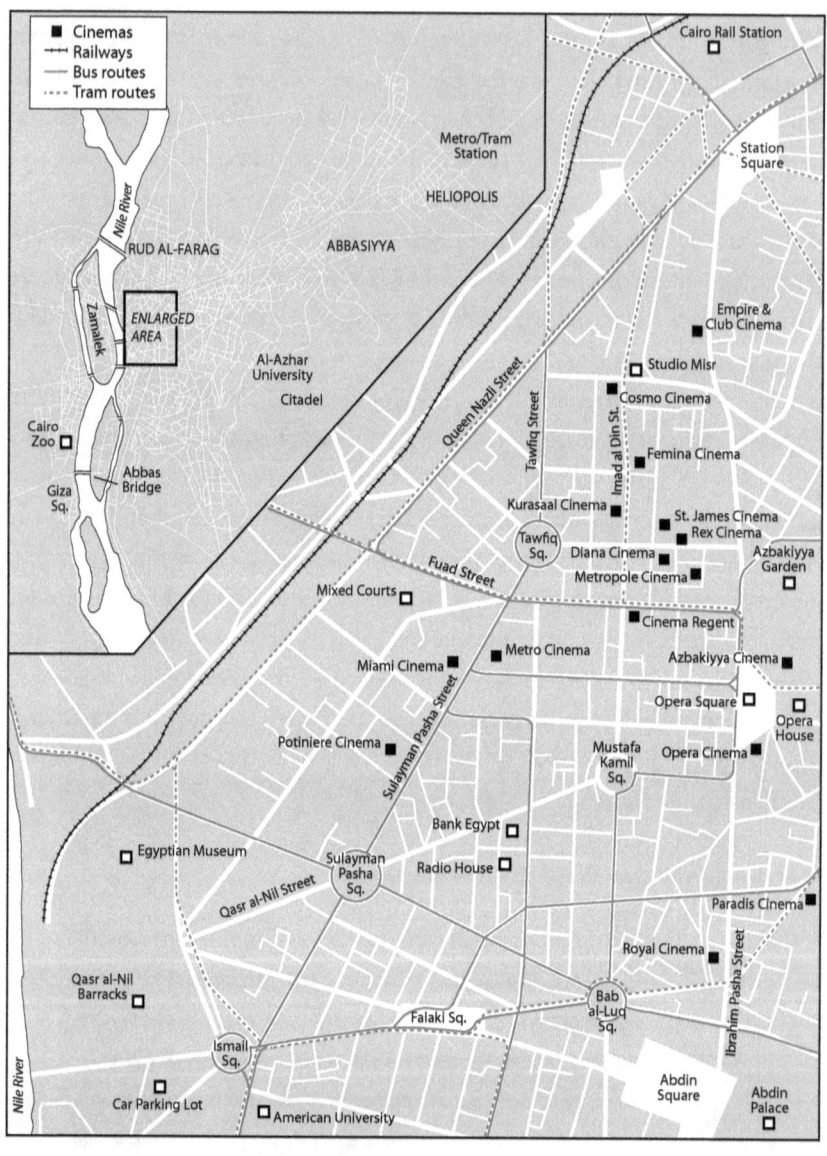

MAP 3.2. Cairo's Tramway and Bus Routes (1944).
SOURCE: Adapted from a street map in the 1944 booklet *Services Guide to Cairo*.

during the first two decades of the twentieth century, car horns were regarded as necessary for alerting pedestrians and avoiding accidents.[72] Not surprisingly though, as the number of motor vehicles increased, car horns in particular were disproportionately adding to the cacophony of the cities and were increasingly viewed as a nuisance in need of regulation. By the mid-1930s, the government drew up anti-honking regulations limiting "noise by instruments of warning" and establishing ticketing and fines for the "excessive use of electric horn or klaxon." However, by the government's own account, these measures—applied in 1936—were not very effective since "fines of P.T. 5 and P.T. 10 did not make drivers respect the law."⁷³ Five to ten piasters for excessively beeping the horn may have been a deterrent for taxi drivers, but for those who could afford their own cars in 1930s Egypt, the fine was simply not enough to curb the noise.

However, efforts to reduce car noise continued in earnest, and on September 10, 1938, a law entirely "forbidding the use of all instruments of warning between the hours of 11:00 p.m. and 7:00 a.m. was put into force." During the first month of enforcing this law, first-time offenders only received letters of warning. Unsurprisingly, this had little impact, and fines were issued for nighttime use of car horns, in addition to the preexisting fines for "excessive use" of horns during the day.⁷⁴ Starting in 1940, the tone of motor vehicle horns was tested and regulated at traffic yards when vehicles were having their licenses renewed. Aside from the concerns about the decibel levels of the horns, these regulations, issued at the start of the Second World War, were primarily concerned with the banning of twin horns, which apparently sounded like air raid sirens.⁷⁵

Creating more traffic laws with noise regulations was one thing, but enforcement was another. Despite the fact that every year thousands of car horn noise violations were issued in Cairo alone, there was little improvement in noise abatement, and this problem was compounded with the steady increase of new motor vehicles entering into circulation every year. Also, throughout the 1930s and 1940s, the Egyptian police—especially their traffic division—had a limited budget and were severely understaffed. As the 1937 annual report of the Cairo City Police succinctly put it, "enforcement of a number of new regulations cannot be effectively carried out unless the numerical strength of the traffic force is increased."⁷⁶

Aside from legal regulations and enforcement of car horn violations, a continuing and steady stream of newspaper editorials and commentary about traffic noise and excessive use of car horns issued from the Egyptian press. With an ever-increasing number of Egyptians attending movie theaters, in 1939 a short public service animated film, sponsored and produced by the Egyptian Royal Automobile Society, used humor to shame Egyptian drivers into limiting their use of car horns. The cartoon, titled *Mafish Mukh* (Brainless), ran for only a minute and a half and featured the cartoon character Mishmish Effendi

(Mr. Apricot), who was already well known to Egyptian cinema audiences (see Figure 3.1). Created and filmed by Frenkel Productions, Mishmish Effendi cartoons regularly played during previews and intermissions in cinemas throughout Egypt.[77] The new cartoon effectively made its anti-noise message through humor and vernacular dialogue, voiced by comedian and monologist Ahmad Mitwali. After showing and loudly demonstrating the noises of Cairene traffic, the film introduces Mishmish driving his car and stopping at a red traffic light. Mishmish is visibly disturbed as an impatient driver behind him incessantly honks his car horn. Mishmish then speaks up to the honking driver: "What's the use of beeping the horn? There are no pedestrians walking in front of you!" When the driver does not respond and continues to honk his horn, the exacerbated Mishmish yells out, "Man . . . can't you tell that it's a red light and not a person that you're beeping at!" At the end of the cartoon a written message appears: "Do not use the car horn unnecessarily!"[78]

Vital to the messaging of this short cartoon is the verbalization, pronunciation, and tone of Mishmish as contrasted with that of the loud driver. There is

FIGURE 3.1. Mishmish Afandi and Traffic Noise.
SOURCE: *Mafish Mukh*, Frenkel Brothers, 1939.

a clear class and educational distinction represented in the ways they speak the Egyptian vernacular. Although both have a Cairene accent, Mishmish clearly sounds like an educated, "modern" effendi, whereas the man honking the horn responds to Mishmish in a stereotypically lowbrow Cairene Arabic. This mediated lower-class voice as spoken by the loud, car-horn-beeping character associates loudness, yet again, with classless or "lower-class" behavior. It is unlikely that the Mishmish Effendi cartoon had much of an effect on the ubiquity of car horn noises in Egypt's urban streets, yet it highlights the changing sonic realities of cities in the 1930s, while revealing changes in middle-class sensibilities and discourses about noise. In any case, as I examine next, all the varied cacophonous sounds produced by the ever-growing urban traffic, from motor vehicles, trams, and animal-powered carriages, came to a head in some of the key squares and transportation hubs of Cairo and Alexandria.

The Hubbub and Sensory-Scape of Squares, Stations, and Transport Hubs

> The Ataba-el-Khadra, in the angle of the Mixed Tribunals and the Post Office, is the best place to observe another kind of native life. The Arabs are extremely fond of using tramways and omnibuses and take them as seriously as we take catching a train. As they are bustling-in they are waited on by a swarm of vendors of tartlets, Turkish delight, seditious newspapers, and tinkery and turnery, not to mention the swarm of water-sellers, lemonade-sellers and shoeblacks, or the donkey-boys and the *arabeah*-drivers [carriage drivers], who deafen you with their noise, and the forage camels and stone-carts who jostle into everybody.
>
> Douglas Sladen, *Oriental Cairo* (1907)[79]

One of the main reasons for the Haussmannization of cities in the nineteenth century was to better control the urban masses by making it easier for the state and its various institutions, from medical vaccination officials to police and taxation officials, to control hard-to-reach areas in the heart of the city. Ironically, these very same large squares and wide streets allowed the urban masses to also more easily commute, move, and work in the heart of the modern city. In times of national or political crisis, these new wider spaces gave the urban masses unprecedented room to congregate, meet, and demonstrate en masse.

The Egyptian Revolution of 1919 would have been unimaginable without the wide streets accommodating thousands of demonstrators in long marches and, more importantly, without the large squares that allowed tens of thousands to stand shoulder to shoulder as a diverse "national" crowd, collectively experiencing (albeit briefly) a literal multisensory embodiment of the various grievances and hopes of the nation. Or, looking at more recent events, it is almost impossible to conceive the 2011 Revolution without Tahrir Square.

Large and modern city squares are an important and qualitatively different socializing space from traditional city quarters. In mundane, everyday settings, city squares are vital nodes of interaction where large commercial, social, and transportation activities daily take place. With the advent of trains in the mid-nineteenth century and tramways a few decades later, train and tramway stations, and the areas immediately surrounding them, became loud, huge hotbeds of economic and social activity. The hubbub was often punctuated either with rush hour in the case of the tram or with the sudden spurt of activity regularly taking place with the arrival of a train.[80] In Cairo, large central train stations like Cairo Station in Bab al-Hadid Square and tramway hubs like al-Ataba al-Khadra Square became cacophonous mega economic and social centers where everyday people traveled or daily commuted to work[81] (see Figures 3.2 and 3.3). Coaches, transportation carts, horse-drawn omnibuses, and later on taxis and buses assembled en masse in and around these stations and squares to pick up and drop off train and tramway passengers.[82] Naturally, these very same spaces drew in large numbers of commercial activities. Thousands of merchants, tourist guides, food vendors, newspaper boys, shoeblacks, street hawkers, luggage porters, beggars, and others would congregate near and around these transportation centers to sell their services or their goods and wares. As I examine in more detail in the next chapter, department stores, bars, cafés, travel agencies, theaters, cabarets, and other business also strategically opened up shop near some of these mega squares and stations.

Al-Ataba al-Khadra Square was the primary transportation hub for the entire city of Cairo, and imagining its sonic transition over time is a useful exercise. At the heart of downtown Cairo, Ataba Square was the nerve center of all the multiple transportation networks intersecting the city. As the major junction for all tramway lines in greater Cairo, the square was constantly crisscrossed by tramway traffic connecting passengers to every corner of the city. Also, it was at the intersection of all the major roads connecting the more densely populated

FIGURE 3.2. Al-Ataba al-Khadra Square in 1907.
SOURCE: Photograph by Max H. Rudmann, Egypt 1907.

parts of old Cairo on one end and the roads leading to the newer districts on the other end. During the trams' hours of operation from 5:00 a.m. to 1:00 a.m., wheeled traffic from cars and buses to carriages and carts was regularly arriving to and leaving from the square, bringing and carrying tramway passengers almost around the clock (as seen in Figures 3.2 and 3.3). On top of all this, the hustle and bustle of al-Ataba al-Khadra Square was no doubt augmented by the hundreds of street hawkers, calling out and selling everything from knickknacks and newspaper, to all kinds of food and drink.[83]

Station Square rivaled Ataba Square in its sensory chaos, as it housed both Cairo Station, the most important train terminus in Egypt, and also, right outside the train station, al-Laymun Bridge Tramway station, one of the critical

FIGURE 3.3. Al-Ataba al-Khadra Square in 1947.
SOURCE: Bibliothèque nationale de France, Département des estampes et photographie, EI-182, http://catalogue.bnf.fr/ark:/12148/cb44395799f.

tramway stations in Cairo (see Figures 3.4 and 3.5). Almost all forms of ground transportation, including motorized buses and taxis as the twentieth century roared on, congregated in large numbers in this major square, catering to the constant arrival and departure of train and tram passengers.

The sensory-scape of this environment must have been overwhelming to the many new arrivals daily coming in from the countryside. In fact, one of the reoccurring cultural tropes that captures the overwhelming visual and auditory cacophony of Cairo Station is the continual retelling of the narrative of a peasant (usually from Upper Egypt) arriving in Cairo from the countryside for the first time. The peasant is typically portrayed as being flush with cash, having just sold his land or the season's cotton harvest. Upon arrival at Cairo Station, the peasant is invariably awestruck by the sights and sounds of Cairo and almost immediately approached by a Cairene conman who takes advantage of his disorientation. The result of this interaction leads to the fellah losing all of his money by buying from the conman either the station's clock tower, a train, a bridge, or some other building in Cairo. You can still hear variations of this story as a joke in Egypt today, and traces of this basic plotline were and still are depicted in Egyptian plays, cartoons, and movies.[84] This narrative is not unique to Egypt, as stories of scams to sell or rent the Brooklyn Bridge after it was built in 1883 are still part of the American lexicon. These stories, of course, have some truth to them, as many Egyptian peasants and others were inevitably pickpocketed in Cairo Station upon arriving in the city, and acts of fraud were undoubtedly perpetuated against some landowners who arrived at the capital searching for investment opportunities.

FIGURE 3.4. Station Square: Outside Cairo Railway Station in 1900.
SOURCE: U.S. Library of Congress.

FIGURE 3.5. Station Square (ca. 1950).
SOURCE: Postcard photograph of Cairo Station Square.

More importantly, such narratives, whether recounted in Brooklyn or Cairo, reflect people's anxieties about the rapid modernization taking place and also echo the common alienation resulting from large-scale migration and urbanization. Also, the unaccustomed anonymity that peasants from small villages must have felt as they adjusted to living in a metropolis must have been a jarring experience.[85] In 1958, Cairo Station was immortalized in *Bab al-Hadid*, a classic Youssef Chahine movie that depicts the lives of baggage porters, paperboys, street hawkers, soda sellers, and others who depend on the train station to make a living.[86] The film was also much concerned with social apprehensions about modernity and rapid urbanization, particularly among recent migrants from the countryside.[87]

The Electric Tramway as an Embodied Socioeconomic Space

> Like every technology, the electric streetcar implied several businesses, opened new social agendas, and raised political questions. It was not a thing in isolation, but an open-ended set of problems and possibilities.
>
> David Nye, *Electrifying America*[88]

Reliable and affordable tramways and, later on, public buses dramatically expanded the circulation of people in Cairo and other Egyptian cities. However, electric tramways were not just vehicles for transporting people; they also created their own social space, allowing for direct and indirect sociability and interaction across gender, culture, and class lines. In Cairo, Alexandria, and Port Said, only the upper classes and those who could afford their own private automobiles or carriages could altogether avoid taking the tram.[89] Egypt's working classes and middle classes, and also a wide spectrum of non-Egyptians living in Egypt, frequently used tram services for their daily and nightly commute. Placing such a wide range of social classes and people together in a semi-confined space on a daily basis was a new phenomenon, and it had many social implications. To be sure, there were separate areas dividing the tram into first- and second-class sections, though, as mentioned, the difference in ticket prices was not vast, even after they had increased to five and ten millimes for second and first class, respectively, by 1912.[90]

As discussed earlier, there was also a separate first-class women's section, though women could also sit in the non-sexually segregated areas of both first and second class. Although the open car trams, which predominated in Cairo, allowed for constant passenger crossings between these sections, so there was never a regular way to enforce these separations. Providing a semi-segregated space only to upper-middle-class and elite women, who represented a miniscule percentage of the Egyptian population, while appearing to be a contradiction was in fact consistent with cultural norms at the turn of the century.[91] As we have seen earlier in this chapter, the vast majority of Egyptian women, who were peasants or poor and working-class urbanites, have always had significantly more presence in public life than upper-middle-class and elite women, as they regularly interacted with other men and women in the fields, markets, workshops, factories, and streets.[92] Tramway space was simply an extension of these other spaces, but it also allowed many middle- and upper-middle-class women to be a part of that same "public" space. But how public is a tramway car compartment as opposed to, let's say, walking in the street?

James Ryan's examination of tramways in Istanbul, using an adaptation of Lauren Berlant's "intimate public sphere" concept, can bear fruit when used to examine tramways in Egypt.[93] In Cairo and Alexandria, there was ample evidence that tramways provided a vehicle for "intimacy" and gender mixing among all classes of Egyptians in the early twentieth century. From the mid-1890s to the 1930s, when visible public gender mixing by Egypt's middle and upper-middle classes was still viewed unfavorably, the Egyptian press was full of alarming editorials over the decline of morality. As On Barak has already shown, Egyptian tramways were considered by some as one of the primary culprits facilitating these unwanted public interactions among the sexes.[94] The "intimacies" of tramways and public transportation in general extended to social and economic interactions. Social interactions began well before the tram even arrived, as people waited in close spaces next to each other at stations and stops interspersed throughout the city. The proximity of the social and class intermixing taking place in such small quarters inevitably made it embodied, in the sense that most of the senses had be consciously or subconsciously engaged. Crowded tramways necessitated people sitting and standing shoulder to shoulder. Those who were standing likely reached out to grasp the metal bars overhead, lest they fall or bump even further onto the other passengers. Complementing the tactile interactions, passengers engaged in small talk, overheard conversations, and emitted or smelled a variety of pleasant or unpleasant artificial scents and natural odors. Tramways were not hermetically sealed capsules, and sensory perception and engagement with the street was ongoing for all tramway passengers, especially for those sitting or standing near the open doors and windows. In fact, most Cairene trams, save for a metal side rail, were entirely open to the street (see Figure 3.6).

Embodiment in the Tram: From Tram Surfing to Pickpocketing

As alluded to earlier, tramways in Egypt provided a potentially liminal space of interaction that at times was not entirely public. Whether the tram was moving or not, and how fast or slow it moved, contributed to the degree of spatial isolation experienced by the tram passengers. The faster it moved, the more isolated and semiprivate its riders would feel as they raced through the city streets. On the rare occasions when the tram sped up, the separation from the outside increased as it became progressively harder for passengers to leave or enter the tram's space, and the sensory perception of the outside world became increasingly blurred. Most of the time however, the trams traveled at relatively slow speeds, especially since many of their tracks were laid in the middle of the street and the trams had to avoid colliding with vehicular and pedestrian traffic.[95]

FIGURE 3.6. Crowded Cairene Open Tram (ca. 1961).
SOURCE: Eugene Harris Collection, American Geographical Society Library, University of Wisconsin-Milwaukee Libraries.

The time of day or night and the number of passengers in the tram also contributed to how private or public the tram felt to its riders. This liminal effect also extended to the fact that the line between pedestrian and tram rider was not always clear, as it was easy for paying and even nonpaying individuals to go in and out of the tram when it stopped in traffic or was moving at slow speeds. Not only were the cheapest tramway tickets affordable to most, but by avoiding the conductor altogether, many could easily use the tram without paying a penny. This was frequently done by children, students, and street merchants who could easily climb in and out of a tram or simply hang onto the outside of the moving tram and jump out to the street when approached by the conductor. Getting on and off a moving tram required quite a bit of physical and mental dexterity, of course, coupled with an instinctive sensory awareness of various street obstacles and an instantaneous calculation of the variable size and speed of oncoming motor and animal traffic, as well as bicyclists and pedestrians.[96]

Multilingual signs in Arabic, English, French, and Greek were posted on the trams that forbade passengers from riding on the footboard running outside the tram.[97] These signs were rarely obeyed in practice, as footboards were almost always in use by paying customers when the tram was overcrowded and by tram surfers who latched onto the outside to avoid the fare. Newspaper boys, shoeblacks, and sellers of small trinkets also relied on the footboard for jumping on and off the moving tram at will. This sort of tram surfing was very common in Egypt, as this 1918 account of street hawkers using this method attests: "If the regulations about riding on footboards were enforced the hawkers of meats and drinks and curios [knickknacks] would not plague you with their constant solicitation. The boot-boys carry on their trade furtively between the seats: often they ride a mile, working hard at a half-dozen boots."[98]

There were ongoing attempts by the police to limit tram surfing, and in 1921, just in the 'Abdin district of Cairo, 731 citations were given for "illegal tram riding." However, by all accounts, this was just a drop in the bucket and did little to discourage the practice.[99] For example, an American couple living in Egypt in 1955 remarked: "We were able to spend many enjoyable hours exploring the city. These journeys were usually made by tram, a most exciting experience, as the majority of Cairo trams are open at either side, and travel so slowly that peddlers can come and go at will, even when the tram is in motion."[100] Paperboys in particular but also sellers of small trinkets specialized in going into and out of trams as they stopped in stations, but quite often they could also easily enter and exit the tram while it was traveling at slow speeds in the busy urban traffic. To maximize the speed of their transactions and to attract more buyers, they would of course call out loudly, advertising their newspapers and their wares. For newspaper boys, calling out or singing the titles of the magazines or newspapers was sufficient. If there was important political news, they sometimes shouted out the headline of the day.

The liminality of the tram as a space, its often overcrowded conditions, and its porousness to the streets also allowed and facilitated pickpocketing, as a 1920 Ministry of the Interior report relates: "The open trams enable pickpockets to jump on and off the carriages and a large proportion of thefts are committed on passengers in these public vehicles."[101] In every sense, more so than even tram surfing, pickpocketing is an embodied and intimate experience. Thieves had to have all their senses operating at maximum capacity in order to touch their victims and steal their wallets or personal effects without being perceived. For the theft to be carried out successfully, the victim had to be in a crowded area, standing or sitting almost shoulder to shoulder with others. This constant physical touching desensitizes the victim to the deft and light touch of the thief. An accomplished pickpocket could bump into the victim "accidently" while simultaneously swiping a watch, wallet, or jewelry. This accidental bumping is facilitated in a moving tram or bus, where bodies can sway from the changing speed and direction of the vehicle. Sometimes pickpocketing is not a solitary act, as two or more thieves might cooperate so that one distracts or bumps into the victim while the other swipes the money or jewelry. Though difficult to statistically corroborate, the same Ministry of the Interior report claimed that the "flowing robes worn by a large proportion of the public enable the thief to insert his hand in the pockets without alarming the victim."[102] This is certainly possible, though this description might also reflect the writer's class or cultural bias against Egyptians who wore traditional clothing.

In any case, as we have seen, public transportation and trams in particular provided a moving platform for riders to talk, buy, sell, and engage in a variety of social and economic interactions. As a physically moving space that "outsiders" can enter and exit with ease, tramways are neither private nor entirely public. In many ways, the activities of tram surfing and pickpocketing illustrate the relative liminality of the tram as a new modern space that on a daily basis allowed for socially mixed and loud embodied interactions. Moreover, the use and "misuse" of the trams by paying and nonpaying passengers, street hawkers, and pickpockets indicates that trams were for the most part almost entirely appropriated by the urban residents for their own purposes, going well beyond their intended use for basic transportation.

Conclusion

The rapid worldwide introduction of new transportation technologies from the mid-nineteenth to the mid-twentieth centuries brought unprecedented and dramatic change to the soundscapes of most inhabited areas. That was certainly the case in Egypt as rapid growth in vehicular traffic pressured the infrastructural limits of Egypt's roads and forever changed the ways in which urbanites sensorially experienced public space. This chapter considered the burgeoning of the road and track infrastructure in Egypt along with the rapid and successive introduction of animal-powered wheeled vehicles, motor vehicles, and the electric tram. For a time, these new electric and motor-powered vehicles coexisted with animal-powered carts. However, gradually and over the decades, animal power decreased its hold on Egypt's transportation sector.

The level of social, economic, and sensory interaction in large city squares, in particular, was unprecedented, as people physically experienced and negotiated these new transportation vehicles and thousands of other people on a daily basis. After large city squares were built, they were one of the few places where men and women of all classes daily interacted in the same space. Unlike the traditional marketplaces in the heart of old Cairo, the sheer scale of these modern squares provided an altogether different sensory experience. Depending on where a person was sitting or standing, the field of vision of that individual could capture hundreds or, in a rush hour, even thousands of other individuals carrying on with their daily lives. The multiple and varied sounds of all of these activities could also be heard. The sensory-scape of these new larger spaces produced an entirely different experience from that of the more traditional, enclosed urban quarters, where residents knew each other by face, voice, and most likely even name. Adding to the auditory and visual disorientation and the complete anonymity of these large crowded squares, were a plethora of smells from engines, gasoline, and various vended foods and drinks, and the ever-present aroma of horse and donkey dung that inevitably filled the air. If one were riding in a busy tram or bus, physically touching other passengers would complete what must have felt like a continual sensory bombardment.

The rapid growth in vehicular and especially motorized traffic overstretched the infrastructural limits of Egypt's roads and dramatically changed how Egyptians commuted and experienced their immediate environment. Trams and buses in particular, with their speed, sounds, and smells and their ability to simultaneously transport dozens of diverse men and women in a semi-enclosed space, forever changed the way many people sensorially experienced public space. Buses and trams also affected pedestrians and also machine-, human-, and animal-powered vehicles, as everyone had to visibly and sonically project themselves and/or their vehicles in order to navigate the increasing complexity of city traffic. The extent of socioeconomic and sensory interaction in these city streets, large squares, and transportation hubs was unprecedented, as men and women physically experienced and negotiated these new transportation technologies and thousands of other people on a daily basis. These dynamics are generally overlooked in the historical literature; yet by listening to the sources and paying attention to the sensory impact of these transformations in the street, we cannot help but get closer to the mundane realities of everyday life.

4

The Soundscapes of Modernity
Electricity, Lights, and the Sounds of Nightlife

> There are two clear scenes of Salama Street [in Sayyida Zaynab District, Cairo] which remain in my memory quiver [from childhood], both are filled with light. The first scene took place the night when the Tanzim Ministry introduced gaslight to our street; lighting it with a bright white light, the likes of which we had never seen before. We stayed up late that night underneath the streetlight until dawn. . . . Although in our house we already had electrical light for a few years before this night, experiencing the bright light outside in our own street, produced a sense of collective euphoria.
>
> Fathi Radwan, *Khati al-Ataba*[1]

The introduction and spread of electricity starting at the turn of the twentieth century brought with it abundant sources of light, power, broadcasting, amplification, and locomotion, which would soon be felt, albeit unevenly, in Egypt's major urban areas. Starting in the cities and large towns in the late nineteenth century, electrification gradually spread electric light, supplementing the already established modern gaslight that had been introduced by foreign-owned private utilities just a few decades before. In a pattern similar to that appearing in the rest of the world, there was a clear disjunction between economic, cultural, and even social realities before electricity and after.[2] As Fathi Radwan makes clear, municipal gaslight and, later on, electric streetlights meant more consistent human activity well into the evening and nighttime hours, leading to more sound and noise. Electricity, with all of its varied offshoot technologies from lights to telephones and tramways, drastically changed how the world sounded, moved, communicated, and socialized.

This chapter continues our examination of roads, tracks, and modern public spaces in Egypt, by expanding the focus to the professional lighting of Egyptian towns and cities, the growth of a regular nightlife, and the establishment of newer places of public leisure. The sonic implication of electricity, whether relating to night lighting or the gradual introduction of many sounded electrical machines and technologies, from radios and movie theaters to loudspeakers, was enormous. The extension of mundane activities well into the night, changed the sleeping, leisure, and consumption habits of hundreds of thousands and in the process made the nighttime louder than it has ever been.

This chapter will briefly account for the modern lighting and electrification of Egypt during the first quarter of the twentieth century. In the process, I will survey some of the new urban public spaces in Egypt, especially those that specialized in providing leisure activities for everyday people. These included more traditional markets and cafés—which by the 1930s increasingly had blaring radios and electric lights—to newer and well-lit entertainment and shopping districts. More importantly perhaps, I will take into account the electrical transformation of Egypt's boisterous nightlife and the creation of regular entertainment districts catering to diverse audiences. The class implications of these sonic transformations were significant. As nighttime noise levels from blaring radios projecting from working-class Egyptian cafés increased, so did the complaints, which flooded the press. As I examine later, in actions similar to the anti-noise legislation that targeted the calls of street hawkers and car horns, the government also attempted to regulate the public use of radio sets.

Lighting the Night: From Gaslight to Electricity

> To his Excellency the Prime Minster of Egypt,
>
> We the owners of the barbershops of Cairo are asking your eminence to stop the enforcement of the recent ordinance ordering us to turn off our lights at 7:00 p.m. As your Excellency is probably aware, our busiest work is conducted during these limited night hours, as we are dependent [*murtabitun*] on servicing employees and workers such as, lawyers, doctors, and various craftsmen, who cannot use our facilities until well after this limited time (i.e., 7:00 p.m.). Also, the government headed by your Excellency, in all of its wisdom has allowed cafés, retail stores,

> restaurants and pharmacies to keep their lights on until 10:00
> p.m. . . . Although the government now allows us to light our
> stores using candles and oil, both of these are too expensive and
> consume most of our profits. Also, the poor light they produce
> makes it impossible for any barber to properly perform their job.
> We therefore beseech your Excellency, to allow us to light our
> stores until 9:00 p.m. . . . each according to what they can afford.
>
> Petition by Cairo's Barbers (1917)[3]

The petition displayed here, written by a group of over two dozen Cairene barbers, was sent to the Egyptian Prime Minister on August 21, 1917, in protest of the British wartime energy policy (applied in the summer of 1917) that forcibly shut down electric light and gaslight in many of Egypt's urban areas. This petition was part of a coordinated national effort, as I located another very similar petition that had been sent three days earlier by the barbers of Alexandria.[4] What especially incensed the barbers was that most stores and restaurants were allowed to keep their lights on until 10:00 p.m., but barbershops were obligated to darken their shops by seven in the evening. More specifically, it seems that the government banned the barbershops from using electric light and municipal gaslight but allowed them to use "candle and oil lamps." The petitioners balked at this, however, rightly pointing out that the inflationary high price of these items would practically bankrupt them. More importantly, they specifically mentioned that the low quality of these sources of light was not sufficient for them to run their businesses. These austerity policies were part of many other rationing measures imposed by the British during the First World War. Since almost all of Egypt's electric power plants ran on either oil or coal, which were in short supply during the war, electricity consumption had to be drastically cut down.[5]

The Cairo barbers' petition was signed by about three dozen barbers from several districts in Cairo, though a majority were from the reasonably well-to-do and "modern" upper-middle-class areas. Some of the signees wrote their street address next to their signature, and quite a few were from 'Abdin district, Clot Bey Street, and 'Abd Al-Aziz Street, which at the time were at the heart of modern Cairo. This, in part, explains the availability of electric light in their shops in 1917.[6] Indeed, before the 1917 lighting restrictions, contemporary accounts describe the streets of Cairo in and around the Azbakiyya district as "brilliantly lit up" and the town and harbor of Alexandria as "cheerfully illuminated . . . after dark."[7] In these modern districts, electric lights were by this time so prevalent that merchants and shop owners depended on them for their day-to-day operations and for attracting customers.

These petitions corroborate the ways in which electric and municipal gas lighting became indispensable for businesses, as they dramatically extended regular hours of operation well into the night. Asking to extend their lighting hours for just a couple of hours more, until 9:00 p.m., revealed an economic necessity for barbershops, and most likely other stores as well. Electric light was not needed just to supply enough light to see at night and operate the store or business; bright lights also attracted customers like moths to a flame. By the late 1920s, the use of well-lit glass storefronts on department stores, displaying clothes on mannequins and the other goods, began to spread from the wealthy areas of Cairo and Alexandria to middle-class districts and beyond.[8] As soon as the electrical wires and infrastructure reached a district, cafés, restaurants, and shop owners undoubtedly made use of electric light to attract more customers and extend their hours of operation into the night. This was only the beginning, of course, as electricity was becoming cheaper and improvements in electric lighting continued throughout the 1930s and 1940s. Wherever nighttime lights spread, so did nightlife, nighttime conversation, and sounds and noise.

Municipal Lights

Before proceeding further in our analysis, it is vital to briefly overview the history of gas and electric lighting in Egypt. Knowledge of some of the basic technological innovations and key terminologies will provide context and a better understanding of the economic, social, and cultural forces at play. Putting aside the powerfully bright arc lights, with luminosity measured in thousands of candlepower, in the late nineteenth century and well into the first decade of the twentieth century, gaslight and the incandescent electric lightbulb were almost interchangeable in brightness and quality of light.[9] It was later advancements in lamp filament technology that increased the brightness of the electric lightbulb and, in short order, made it indispensable for lighting up the rest of the twentieth century. Edison's original filament of choice, carbonized bamboo fiber, was sufficient but not quite good enough. Just before the First World War, however, the tungsten incandescent lamp was invented, which transformed the weaker light of the turn-of-the-century bulb to the "blinding white light of a modern 300-watt light bulb."[10]

Modern gas lighting was introduced into Egypt in the mid-1860s, and electric light in the mid-1890s. By the end of the nineteenth century, electric light was being gradually introduced in some government buildings and institutions and in the more well-to-do districts of Cairo, Alexandria, and the new canal cities. By the second and third decades of the twentieth century, electric streetlights started to crowd out gaslight in some of the key commercial and social centers of Egypt's northern cities, though gaslight continued to be an important part of street lighting well into the 1940s. Just like the previous municipal gas lighting process, the electrification process was completely decentralized and quite uneven, and it was foreign-owned utility companies that began providing electricity to government buildings, some major urban streets, and wealthy households by the late nineteenth century. As a rule, the newly built urban areas where most of the well-to-do Egyptians and foreigners lived were able to get the benefits of electricity first, but as the technology improved and the price per kilowatt-hour of electricity decreased, starting in the late 1920s, more and more Egyptians gained access.

Some of the private foreign companies, like the French Lebon Company—originally a gas lighting company—were already supplying gaslight to some of Cairo's and Alexandria's main streets and buildings, and they had to quickly transition to providing electricity and electric light as well. More electric power plants were built by Lebon and other foreign utilities, as demand for electric power for factories, street lighting, and tramways was increasing.[11] In a manner of speaking, the Lebon Company was also competing with itself, as its electric light division gradually made its gaslight division obsolete. The company nonetheless fought tooth and nail to keep its original gas lighting contracts with the Egyptian government. In its revised 1893, 1905, 1909, and 1922 contracts, it vigorously insisted on keeping its gas lighting infrastructure and would add electric street lighting only to new areas not already served by its gaslights.[12] The revisions were thus mere addendums for providing electric lighting, added to the initial 1865 gaslight concession almost as an afterthought.

Potential competition from a variety of sources would soon complicate the power supply market. Some private and government-owned factories built their own power plants to supply their own factories with electric power for machinery and light. The Egyptian Government Railway, the Delta Railway Company, and even water utility companies built their own electric power plants.[13] Tramway companies also built their own power plants, not only to supply power to the newly introduced electric trams but also to light the streets used by their tramway cars and tracks, for safe nighttime operation. Each of the five tramway companies in Egypt, located in Cairo, Alexandria. and Port Said, had its own electric power plant. Competition was fierce between private power companies like Lebon and the tramway companies, who by the early twentieth century already had growing tramway networks in Alexandria, Cairo, Heliopolis, and Port Said. The protectionist instincts of the Lebon Corporation in this new competitive environment can clearly be seen in its frequently revised contracts and its correspondence with the Egyptian government. For example, in Article 10 of the revised contract with the government, Lebon grudgingly accepted that tramway companies were to be allowed to "light up the streets, places and avenues in which the tramway cars circulate." However, the article added, "it is understood that the lighting will supplement and not replace the gas lighting that may already be in place."[14]

Thus, by the 1910s, there were, in effect, several separate electric grids provided by competing and diverse institutions and companies. This decentralization continued to a large extent until the late 1940s, prompting Robert Vitalis to accurately label Egypt's electric infrastructure at that time as a "crazy-quilt of public and private power plants."[15] This general chaos and decentralization was reflected among all the utility companies, leading in the 1930s to popular calls for government control of all utilities. These demands, of course, were also intensifying for nationalist reasons, increasing pressure on all the foreign utilities, including Lebon. The allegations leveled at these foreign corporations by nationalists and the Egyptian press were that they were exploitative and charged unreasonable prices.

Despite the obstacles, costs, and bureaucratic difficulties standing in the way of electrification, the broader yet gradual trend was toward cheaper prices and the increasing availability of electricity. From 1921 to 1922 alone, the number of private consumers of electricity increased dramatically, and as coal and oil prices began to moderate sharply, the cost of electricity was reduced from 44 millimes to 34.6 millimes per kilowatt-hour (see Table. 4.1).[16] By the late 1920s, prices had been reduced by almost 30%, yet were still high by global standards, which created a great deal of controversy, negotiation, and debate throughout the early 1930s.[17] In the process of these long negotiations, it was leaked to the public that through currency manipulation and other means, Lebon was keeping electricity prices in Egypt artificially high, and had been doing so from the very beginning.[18] For good reason, resentment of such exploitation by foreign private utility companies was growing in Egypt, and as Robert Vitalis has shown, this would eventually make "nationalization and extension of the power sector a priority for post-war reform."[19]

In Cairo, Lebon's concession was terminated at the end of 1947, and the Egyptian government "took over all the company's power stations and installations in the Cairo District."[20] Lebon had been unable to expand power production or build new power plants during the Second World War. Therefore, the first order of business for the newly formed Cairo Electricity and Gas Administration (CEGA) was to dramatically expand the output of Cairo's electric power plants. Working toward this end, in 1948 the government authorized over 6.5 million Egyptian pounds to build a new power station in north Cairo.[21] Electrification would continue to expand in Egyptian cities and large towns, reaching more and more neighborhoods and households in the post–Second World War period. The high stakes intrigue and corruption associated with the 1930s Lebon negotiations, when coupled with Egypt's dependence on imported coal, explains, in part, Egypt's push for hydroelectric power in the next few decades, including perhaps President Gamal Abdel Nasser's obsession with building the Aswan High Dam.[22]

Early Domestic Uses of Electricity, 1920–1950

> In truth electricity did not play a major role during my childhood at home. For early on, its major trace was as a simple electric bulb hanging in the middle of the ceiling. For we didn't yet have an electric fridge, a washer, a vacuum cleaner, a fan, air conditioning or a television. Even the radio was at that time considered so precious that we placed it on a high shelf away from prying hands. The electric fridge did not enter our house until 1947 when I was 12 years old. . . . After the refrigerator my father bought an electric washing machine and brought it home without even consulting my mother.
>
> Galal Amin, *Madha 'Alamatni al-Haiyyah*[23]

In his memoirs, Galal Amin accurately depicts the gradual exposure of a solidly middle-class Egyptian household to electrification and to increasingly more affordable technologies and appliances. During the first half of the twentieth century, when the price of electricity was relatively high, most households that had access to electricity primarily used it for basic indoor lighting. This was especially true in the 1920s. In 1921 Cairo, for example, the Lebon Company provided indoor electric lighting to 16,277 households and non-light-related electric power to just 259 households.[24] The electricity consumption of a few bulbs judiciously used to light a house at night was fairly low, and lightbulbs were inexpensive. Lightbulbs became even less expensive in the 1930s when they were manufactured locally in Alexandria and Ismailia. By the late 1930s, local bulb prices not only undercut the imports but substantially reduced the overall number of imported lightbulbs.[25] According to British economic reports, in 1946, a single newly built Egyptian factory was producing a million and a half bulbs a year and had the potential capacity to manufacture "approximately 3 million lamps per year," which was the estimate for Egypt's entire annual consumption.[26]

Though the supply of electricity was increasing in Egypt in the late 1930s, electricity prices were still high, which did not encourage the buying and use of large electric domestic household items. Electric refrigerators for example did not become regularly attainable for most middle-class households until well into the 1960s (see Table 4.1).[27] Radios however, not only used less power than refrigerators or washing machines, but as early as the 1930s, a growing number of the radios available in Egypt were battery operated, which naturally made them popular in the countryside and elsewhere where electricity was not yet available.[28] Judging from import reports for electric appliances from the 1930s until the 1950s, radios were the most popular by far, and as their prices decreased in the post–World War II period, they became a must-have electric appliance for any middle-class family. This was, no doubt, not just because of the entertainment value that radios provided but also because they were a visual and sonic display of middle-class upward mobility.[29] On average during the second half of the 1930s, 20,000 radios, worth over 120,000 LE, were imported into Egypt every year. During World War II, the number of imported electric appliances, including radios, dropped to a trickle, only to skyrocket after the war, as prices dramatically deceased and supplies were plentiful. By 1950, for example, 750,000 LE worth of radios were being imported into Egypt annually (see Table 4.1).

These figures do not include the approximately "four to five thousand radio sets per annum" that in the early 1950s were assembled in Egypt using imported parts.[30] Locally assembled radios were much cheaper than the imports, and for that reason, the number of radios increased dramatically as soon as local radio sets became available to consumers. For instance, in 1950, the number of officially licensed radios in Egypt was approximately 264,000, yet in just another

TABLE 4.1. Yearly Imports and Costs of Electrical Appliances in Egypt (1934–1950).

	NUMBER OF RADIOS	COST OF IMPORTED RADIOS (in Egyptian pounds)	COST OF IMPORTED REFRIGERATORS (in Egyptian pounds)
1934	24,847	155,923	—
1935	26,072	176, 246	—
1936	15,267	106,837	—
1937	20,080	125,066	7,831
1938	17,633	107,935	9,961
1947	—	409,504	89,349
1948	—	486,806	267,169
1949	—	582,505	207,983
1950	—	747,024	336,528

SOURCE: G. H. Selous, *Report on Economic and Commercial Conditions in Egypt*, Department of Overseas Trade (London: His Majesty's Stationery Office, 1937), 180; C. Empson, *Report on Economic and Commercial Conditions in Egypt*, Department of Overseas Trade (London: His Majesty's Stationery Office, 1939), 86; A. N. Cumberbatch, Egypt: *Economic and Commercial Conditions in Egypt*, Overseas Economic Surveys series (London: His Majesty's Stationery Office, 1952), 110.

decade over 1.5 million radio sets were publicly registered.[31] Talk and music from radios and ubiquitous sounds emanating from electric refrigerators and fans and later on air conditioners and other appliances, changed the way indoor spaces sounded, not just because of the ambient sounds they produced but also because of the varying degrees to which they sonically isolated household residents by blocking or muffling outside street noise. Indeed, the gradual electrification of Egyptian homes had wide-ranging implications affecting the sleeping, working, entertaining, and studying habits of people. As electric power and lighting became more prevalent, they dramatically extended the regular working and leisure time of Egyptians well into the night. Electric appliances and radios were, of course, not used just in people's homes. In fact, as I examine next, when radio sets were more expensive, from the mid-1920s to the mid-1940s, ordinary Egyptians were most likely to hear songs and other radio programs in cafés and in other public spaces.

Bars, Coffee Shops, and Blaring Music

> Alfresco cafes are ubiquitous. . . . Chairs and tables extend on to the footpaths. The people of all nations lounge there in their fez caps, drinking much, talking more, gambling most of all. Young men from the university abound; much resemble, in their speech and manner, the young men of any other university. They deal in witty criticism of the passengers, but show a readiness in repartee with them of which only an Arab undergraduate is capable.
>
> Hector Dinning, *By-ways on Service* (1918)[32]

In Egypt, coffee shops were, and continue to be, an extension of the street in that their spatial and sonic boundaries almost always expanded beyond their official, "enclosed" space. In theory at least, in the mid-nineteenth century, coffee house owners faced legal prosecution for putting chairs and tables in the main street, though it is difficult to gauge the degree to which these regulations were enforced.[33] By the late nineteenth century, however, a series of laws and legal amendments were issued that allowed registered café owners to pay an annual tax for the placing of tables and chairs on sidewalks and sometimes in the streets. On May 18, 1885, the Majlis al-Nuzzar (Council of Ministers) passed a revised law legislating the proper use of public roads and thoroughfares, which specifically addressed public cafés and bars.[34] For instance, the fourth clause of Article 13 of the 1885 law allowed coffee shop or bar owners to obtain a "permanent long-term license to place tables and chairs in the public streets." The annual license fee was 40 piasters for every square meter. The fee was halved to 20 piasters if the establishment was located in an area "without a paved or a macadamized street." The government, however, gave itself some flexibility by specifying in Article 12 of the same law that "if the owner of the café, bar, or the like, is allowed by the government to place tables and chairs in the sidewalk or road and at a later date it turns out that this impedes traffic, then the government has free rein to either limit this license or to cancel it entirely. If this is done, the owner has no right to ask for compensation."

The original 1885 law was enforced on June 16 in Cairo, Alexandria, Port Said, Ismailia, and Suez. The law was applied in 1887 and 1888 in the towns of Tanta, Zaqaziq, and Mansurah; in 1892 in al-Mahalla al-Kubrah; and in Damanhur in 1894. The towns of Ziftah, Damietta, Banha, al-Minyya, Bani Suef,

Faiyum, Girga, and Qinah enforced the Public Road Usage law in 1896.[35] As with many of these laws, enforcement was at best irregular, and there were always ways to thwart, challenge, and even bribe the authorities in order to undercut actual enforcement. This was especially true if the owner of the café or bar was a foreigner or had protégé legal status from one of the foreign consulates.[36] In Cairo alone, as early as 1887, Ali Pasha Mubarak counted 486 bars, 46 *buza* drinking dens, and 1,067 coffee shops. These numbers had more than doubled by the first quarter of the twentieth century (see Table 4.2). To put these numbers in perspective, in the same survey of Cairene establishments, Ali Mubarak counted only 264 mosques.[37]

At the turn of the twentieth century, there were a variety of establishments in the Egyptian streets that could be broadly defined as cafés.[38] Aside from the hundreds of officially registered and well-established cafés and bars that dotted the Egyptian streets, the traditional hole-in-the-wall cafés were quite popular and primarily served the local residents of their street or quarter. These had much cheaper drinks and fare and were the most prevalent, having existed in some form in Egypt for at least a millennium. According to contemporary accounts, until the 1910s, these cafés usually served more tea than coffee and the tobacco

TABLE 4.2. Cafés, Bars, Restaurants, and Nightclubs in Egypt (1937–1947).

	CAIRO	ALEXANDRIA	PORT SAID & ISMAILIA	SUEZ	TOTAL IN EGYPT
1937					
Cafés and bars	2,711	1,828	516	132	**11,744**
Restaurants	1,473	798	173	66	**5,333**
Entertainment halls (theaters/nightclubs)	36	32	9	6	**108**
1947					
Cafés, bars, and restaurants	4,756	2,649	789	261	**18,584**
Entertainment halls (theaters/nightclubs)	92	36	16	3	**214**

SOURCE: Egyptian Government, *Ministry of Finance: Industrial and Commercial Census*, 1937 (Cairo: al-Matba'a al-Amiriya, 1942), 500, 670, 671; *Annuaire Statistique de l'Egypte 1949–51*, Ministère des finances, Direction de la statistique (Cairo: Imprimerie Nationale, 1953), 885–890.

or hashish was smoked out of a *guzah* pipe.[39] Customers were usually seated on simple wooden benches or cheap, straw-seated chairs. Dominos and playing cards were the games of choice in these small *baladi* (local) cafés.[40] Many of these establishments were quite small and probably not officially registered as cafés by the Egyptian government. Other, even smaller, makeshift coffee shops often popped up at workshops, storefronts, parks, and even barbershops. All it took was a kerosene burner, a few chairs, and a crude table to set up an outdoor space for drinking tea and coffee and socializing. Store owners and managers were always ready to serve beverages to passersby and store clientele, in the hope of encouraging business transactions. Sometimes friends and acquaintances from the neighborhood would stop by, drink, and chat (see Figure 4.1).

On the other side of the café spectrum were the larger Egyptian-owned cafés that catered to government employees, students, journalists, and Egypt's professional classes. These were the cafés that according to Fathi Radwan, served more coffee than tea, and they also, instead of the traditional *guzah* pipe, served tobacco out of a large glass *shisha* (narjilah), or water pipe. Radwan observes that unlike in the smaller *baladi* cafés, backgammon was apparently the game of choice in these establishments, followed by dominoes.[41] These larger cafés

FIGURE 4.1. Two Men Playing Backgammon in a Cairene Café (ca. 1961).
SOURCE: American Geographical Society Library, University of Wisconsin-Milwaukee Libraries.

offered a choice of newspapers for their customers, and were likely able to afford radios when they became available in the late 1920s and early 1930s. The following advertisement, printed in 1906 in *al-Muftah* magazine, describes this sort of Egyptian-owned café and bar:

> Ramsis Bar and Café is located in the healthy and beautiful Fagalla Street. It is owned by the cultured Habib Afandi Ibrahim Bey. The Café has all types of good quality and authentic (not adulterated) refreshments. It's clean and has good service unlike any other café or bar, and is even better than the best foreign owned establishments. We have a wide variety of magazines and newspapers in Arabic and European languages. This makes it equivalent to a literary salon.[42]

As this advertisement reveals, larger Egyptian-owned cafés had to compete with the many foreign-owned cafés. By the late nineteenth century, new foreign-owned cafés, bars, and patisseries—such as Groppi and Café Rich in Cairo and Tiriano, Délices, Athineos, and Pastroudis in Alexandria—were increasingly popular, especially with Egyptian government employees, middle-class Egyptians, and the thousands of Greeks, Italians, Maltese, and Syrians living in Egypt at the time.[43] Cafés of all types were not simply a place for drinking, smoking, and entertainment, as many of them served also as de facto literary, artistic, and political salons. In the theater district of Imad al-Din Street, the appropriately named Qahwat al-Fann (The Arts Café), Qahwat Misr (The Egypt Café), and Qahwat Barun (The Baron Café) were favorite cafés for theater writers, actors, and musicians. In his autobiography, Naguib al-Rihani acknowledges that early in his career, he was practically living in Qahwat al-Fann.[44] In the Egyptian urban sphere, particularly after newspapers became prevalent, starting in the last quarter of the nineteenth century, politics and coffee shops were synonymous. Indeed, as I discuss elsewhere, Cairo's cafés, especially Groppi's, played a loud and crucial role as revolutionary communication hubs during the 1919 Revolution.[45]

Weather permitting, most Egyptian cafés seated more people outside on the sidewalks and on the pavement than inside, with obvious sonic implications for the neighborhoods that surrounded them. This meant that passersby, be they paperboys, street vendors, entertainers, or simply pedestrians, could observe and listen to, if not participate in, the coffee shop experience. Commercial activity was (and still is) loudly conducted on these café sidewalks, as dozens of street hawkers, storytellers, musicians, fortune-tellers, and entertainers took advantage of the already seated and hence captive clientele (see Figure 4.2). Every aspect of

FIGURE 4.2. Outdoor Café with Gas Streetlights (Azbakiyya, near Bristol Hotel) (ca. 1900).
SOURCE: Postcard photograph, by Léon & Lévy, Egypt 1900.

the traditional café experience was sounded aloud and often projected beyond its intended listeners. Typically, orders made in an Egyptian-operated café are not written down but shouted by the waiter to the staff inside. The classic call made by a café waiter to his customers is "*aywa gaay*" (yes, I am coming), which can be heard by all inside and outside the establishment.

Other keynote sounds of a traditional café are the sounds of the typical table games being played. The two most popular games, dominoes and backgammon, have unique sounds that can carry for quite a distance. Traditionally, a domino piece is not placed gently on the wooden table, but is forcefully, swiftly, and loudly clacked on the table, and then with one motion the piece is slid into place, producing a distinct shuffling noise resulting from the pressure the player applies and the friction between the piece and the table. A similar noise is produced in backgammon as the checkers are similarly slid by the players inside the wooden board. Though the noise of the dice as they strike and roll in the backgammon case, producing the classic trick-track sounds, is what clearly differentiates backgammon from dominoes.[46]

Cafés and Booming Radios

Some of the larger, more established cafés were considered *cafés chantants* because they had regular music, dancing, and small theatrical performances.[47] As I examine later, demand for standalone cabarets, theaters, and dancehalls would expand dramatically in the interwar period, creating a robust nightlife.[48] Coffee shops, in contrast, increasingly relied on newer, louder technologies to more cheaply entertain their clientele. Predictably, as electricity was introduced and became increasingly more common in Egypt, the acoustic imprints of coffee shops also dramatically increased. Electricity increased the already considerable sounds emanating from cafés and bars in two ways. Regular electrical illumination, expanded the working hours of these establishments well into the night, attracting more individuals for longer hours. This was especially true for those who still did not have electricity at their residences. Starting in the late 1920s in well-to-do establishments, but expanding rapidly as prices for radio sets and for electric current for night lighting decreased from the 1930s to the 1950s, the use of radios to entertain clients and attract passersby became more and more common. A blaring radio set in a café, workshop, or store might tempt a passerby to stop, listen, and perhaps engage in conversation about the music or the program playing, and then maybe to sit and order something from the establishment.

By the early 1930s, a few years before Egyptian State Radio monopolized the airwaves, there were about a dozen private radio stations broadcasting in Egypt.[49] As already discussed, café owners had started buying radios to attract more customers, and even the more humble establishments were finding a way to buy radios through the payment plans offered by many radio shops and manufacturers, spreading their payments over ten to fourteen months.[50] As early as the summer of 1933, the hubbub produced by blaring radios in cafés and other public establishments was so overwhelming that complaints flooded police stations, government ministries, and the press. *Al-Ahram*, in particular, played a leading role in sensationalizing this issue and advocating for government controls. Throughout the summer of 1933, *al-Ahram* printed a regular column titled "Al-Hukuma wa al-Radiu" (Radio and the government), which supported regulating and in some cases silencing the airwaves. Many of these articles and most of the letters to the editor offered classist definitions of noise, as they typically were up in arms over radios playing in popular coffee shops. Loudly playing the radio, according to *al-Ahram* editorials, was not only disturbing the peace but could lead to public disorder.[51]

In order to control and silence late-night public radio use, the government responded by hastily putting together a two-pronged approach. First, by expanding existing anti-noise law codes to specifically address blaring radio sets, the government could fine and in certain cases even arrest café and restaurant owners for "disturbing the peace by loudly playing their radio set."[52] Or as an *al-Ahram* editorial put it: "Whereas operating a radio loudly in a public space can greatly disturb the peace and restfulness of the general public, it is necessary to limit its use as much as possible. For this reason the government has decided to issue fines and summons to all the owners of various establishments that are open to the general public that loudly use radio without a license."[53] This, of course, would put less well-to-do establishments at a distinct disadvantage.

The second regulatory approach was enacted by the Ministry of Transportation, which was then responsible for all national communications. The ministry arranged a well-publicized meeting with a dozen radio station owners, at which it directed them to limit their hours of operation, especially at night. Photos of the meeting were even printed on *al-Ahram*'s front page.[54] An agreement was put into place as a result of the meeting, limiting the hours of operation for radio broadcasting to "7:00 a.m. to 9:00 a.m., 11:00 a.m. to 3:30 p.m., and 5:00 p.m. to 10:00 p.m. On Thursday, Friday, and Saturday nights broadcasting can continue until midnight."[55]

This restrictive scheduling was immediately resisted by many coffee shop owners, especially those with shops in working-class neighborhoods. According to *al-Ahram*, "many café owners in traditional urban areas wrote petitions to the Ministry [of Transportation] complaining of the new policy of stopping radio broadcasts earlier at night." The café owners specifically complained that this policy would "prevent them from paying their monthly radio payments, as they depend on the customers that stay late in their cafés to listen to the radio."[56] However, there is no clear indication that these new policies were regularly enforced, as just a few days later, a letter to the editor indicated that some radio stations played their programing well beyond the 10:00 p.m. limit on a Sunday. The overly conscientious writer of this letter, alarmingly proclaimed his fears that this "transgression will gradually and chaotically lead to the other stations following suit, defeating the entire purpose of legislating the hours of operation for radio broadcasts."[57] Either way, it would seem that this government intervention did not silence radios in cafés and other public spaces in the short or the long term.

For example, in the early 1940s, the increasing availability of electricity and the transition from storytellers to blaring radio sets in Cairo's cafés was classically depicted in the first few pages of Naguib Mahfouz's novel *Midaq Alley* (*Zuqaq al-Madaq*):

The two houses at the end of the street have closed their shutters against the cold, and lantern light shines through their cracks. Midaq Alley would be completely silent now were it not for Kirsha's café; light streaming from its electric lamps, their wires covered with flies. The café is beginning to fill with customers.... In the café entrance a workman is setting up a secondhand radio on a wall. A few men are scattered about on the couches smoking and drinking tea.[58]

Contrasting the nighttime silence of a traditional Cairo quarter with the electrically lit and soon to be radio-equipped neighborhood café, Mahfouz accurately depicts a transition that was happening everywhere in Egypt. This would accelerate dramatically in the post–World War II era as both electricity and radio sets became more affordable. In fact, by the late 1940s, the prevalence of loud radios in working-class cafés and in other public establishments continued to be frequently commented about in the Egyptian press.

A 1954 cartoon printed in *al-Ithnayn* magazine, for example, satirizes Cairo's growing sound pollution by contrasting the laws that were passed to control motor vehicle horns (see Chapter 3) with the increasing cacophony of blaring radios in cafés and other public spaces. The cartoon caption reads: "Traffic laws regulate the use of car horns so as not to disturb the neighboring residents, but at the same time, radio owners are allowed to loudly blast out their radio sets for an occasion or for no occasion!" (see Figure 4.3). Indeed, starting in the mid-1930s, but becoming more common by the mid-1940s, electric or battery-powered radios became an integral part of Egypt's soundscape. Of course, as I examine later on, the same electric vacuum tube amplification technology that evolved to produce louder radio sets was adapted and used for loudspeakers in theaters, cabarets, public address systems, and sounded motion pictures.[59]

A Regular Nightlife, 1900–1950

> In the Sharia Bab-el-Bahri are the principal Arab theatres, and other places of amusement, and there are always piano organs or bands playing the latest music-hall or comic-opera airs. The whole street is a blaze of electric light. Its ends are taken up with cafés, and its pavements are crowded with vendors of tartlets, sweetmeats, meat on skewers, and sago in teacups; while the cigarette-sellers have stalls that are works of art.
>
> Douglas Sladen, *Oriental Cairo* (1911)[60]

The Soundscapes of Modernity   137

To be sure, in 1909, when Douglas Sladen wrote these words, most of Cairo was still covered in nighttime darkness, forming a stunning contrast to the streets that were a "blaze of electric light" in the Azbakiyya district and in the entertainment districts to the northwest of it. As I examined earlier in this chapter, starting in the late nineteenth century, electrification would unevenly and gradually spread electric light in Egyptian cities, complementing the already existing municipal gaslight and creating for the first time a regular nightlife with all of its entertainment, leisure, commercial, and sonic implications. Wajh al-Birkah Street (later renamed Naguib al-Rihani Street), located just north of Azbakiyya Garden in the Wasʿah quarter, was one of the key red-light districts in Cairo during the first half of the twentieth century.[61] Wajh al-Birkah Street connected Khanzindar Square, where the Bristol

FIGURE 4.3. Radio Noise in the Streets.
SOURCE: *al-Ithnayn*, February 8, 1954.

Hotel was located, to Imad al-Din Street, which by the First World War had become the epicenter of popular theaters and cabarets in Cairo (see Figure 4.4).[62]

Bab al-Bahari Street perpendicularly intersected with the center of Wajh al-Birkah Street, running southward directly toward Azbakiyya Garden. Those three interconnected streets—especially Imad al-Din Street—formed for all intents and purposes, the primary, though not only, entertainment district in Cairo during the interwar period (see Map 4.1). Theaters, casinos, cabarets, bars, brothels, *cafés chantants*, and later on movie theaters were located there in abundance (see Figures 4.4 and 4.5).[63] Accordingly, this wider district, including Azbakiyya, was awash with electric light and provided for a vibrant and raucous nightlife. Similar electrification patterns and vibrant entertainment districts were of course

FIGURE 4.4. Kursaal Nightclub on Imad al-Din Street (1916).
SOURCE: Photograph by Max H. Rudmann, Cairo, Egypt, 1916.

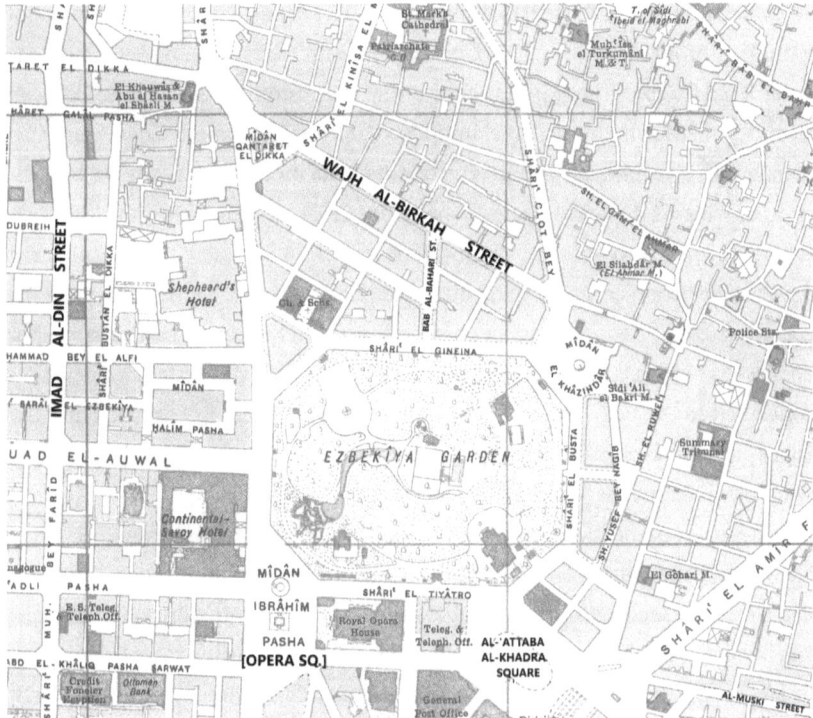

MAP 4.1. Azbakiyya and Cairo's Entertainment District.
SOURCE: Adapted from *Survey of Egypt: Map of Cairo* (Cairo: Maslahit al-Misaha [Survey Department], 1950).

also growing in Alexandria, in Port Said, and to a lesser degree in many of the smaller cities and towns throughout Egypt. For example, despite the lighting restrictions during World War I, memoirs by British and Anzac soldiers and officers abound with descriptions of how Alexandria's "town and harbor . . . were cheerfully illuminated."[64]

There are a few firsthand accounts of some of the nightclubs and dancing halls of Azbakiyya during the first quarter of the twentieth century and a somewhat accurate description of these establishments can be made. The Eldorado, located on Azbakiyya Street, and the Café Egyptien, near Shepherd's Hotel, were two of the most famous *cafés chantants* in the Azbakiyya district.[65] Both featured vaudevillian variety shows, showcasing everything from singers, dancers, and musicians to jugglers, gymnasts, and magicians. Short silent films were sometimes shown between acts. Some of these bars and cafés had dedicated stages and performed a variety of plays. Naguib al-Rihani's and Ali al-Kassar's comedic theaters were born during World War I on the stages of the theaters and cafés of the Azbakiyya district.[66]

Most, if not all of these establishments relied primarily on the sale of alcohol to make a profit, as ticket sales for entry were minimal and sometimes customers entered for free. The expectation was that the customers would buy bottles of alcohol either for their own consumption or for the performers. There were also female staff hired almost exclusively to induce the men to buy bottles of beer, liquor, or wine, as the travel writer Lothaire Loewenbach, who visited the Eldorado cabaret in 1907 with two acquaintances, describes: "A Sudanese woman, not very pretty, but with a nice figure, sits at our table. She judges Mr. Wertheim as the most generous one of us and convinces him to buy for her two half-bottles of beer at 10 piasters, which is essentially our entry cover price."[67] This system for payment continued in the belly dancing cabarets of Pyramid Street in Giza well into the 1970s, and the women hired to open the bottles of alcohol were called *fatahat* (sg. *fataha*), which means "openers" in the Egyptian vernacular.[68]

The inside of the Eldorado was described by Loewenbach as being a very large hall with a large stage at one end. He described the typical dance show as having a half a dozen female dancers and singers and an equal number of male musicians playing the oud, table drum, and tambourine. After likening the main belly dancer to La Goulue, a late-nineteenth-century dancer at the Moulin Rouge in Paris, Loewenbach described her pink dress, which revealed her naked belly, as "covered in tacky trinkets." The Egyptian dancer used castanets in addition to the trinkets attached to her dress to sonically highlight her body movement, and paused only to "empty a beer bottle that was sent to her by an admiring spectator."[69] After the show, the belly dancer personally passed around a collection plate, acquiring tips from the audience. The music and belly dance show was the specialty of the Eldorado, and was repeated often during the night. Another part of the entertainment venue at the Eldorado was a silent movie, which was shown after the dance show, and at 1:00 a.m. the dance floor was opened to the public.[70]

Loewenbach's vivid description of the Eldorado dance hall is visually corroborated by a photograph of the electrically illuminated Eldorado stage (see Figure 4.5). The sights he describes—the number of performers on the stage, the dresses worn by the dancers, the musical instruments, and even the castanets on the dancers' fingertips—can clearly be seen. The photograph not only features the musicians, singers, and dancers, but the bottles of alcohol are prominently displayed on the floor of the stage and stacked on small arabesque tables for all to see.[71] The importance of alcohol sales to the owners of these establishments is abundantly clear from the conspicuous displays of these bottles, front and center, along with the singers, dancers, and musicians on the main stage.

Entertainment for the Masses: Loud Summer Nights in Rud al-Farag

> You visit Rod-el-Faraq [Rud al-Farag], the village of cafés-chantants, at night. Only then are the cafés open—and only on summer nights. It is a long ride out by tram through the native quarter. That ride in itself has a fascination. It begins through Boulac.... You will see there what you will see on the further skirts of the city, nearer to Rod-el-Faraq. The native cafés are aflame with lamps and torches.... The haunting music of the single drum and wood-wind strikes the night: the staccato drum and the pipe follow you with a rhythm that persists in your head against the jolting tram. The flat carts of fruit and vegetables are being pushed about over swinging lamps. The vendors are crying their wares stridently and mournfully.... The donkey-carts rattle over the stones, the drivers exhorting the pedestrians, to whom the road is a sort of pavement. The strains of the khan-khan music are emitted from blazing windows above the streets; these are so suggestive that you can visualize the belly dance.
>
> Hector Dinning, *Nile to Aleppo* (1920)[72]

Hector Dinning was an Australian officer stationed in Egypt during World War I, and his observations are rich with a full sensory account of Rud al-Farag's street

FIGURE 4.5. Eldorado Café (ca. 1908).
SOURCE: Postcard photograph in the author's collection.

and nightlife. The loud music is clearly heard in the streets, due in part to the fact that most of the theaters, bars, and cafés located there were open to the air. Moreover, Dinning also relays the sounds of the "jolting tram," the rattling of the carts on the cobblestones, and the sales cries of the vendors. All of these sounds, of course, are the normal sounds of a vibrant nightlife in what was then a peripheral village on the Nile, just north of Bulaq in the northwestern outskirts of Cairo. Before the tram and electricity reached there at the turn of the twentieth century, Rud al-Farag was a small Nile Port village that was, however, abuzz with activity during the day when its vegetable and grain markets were open. The advent of the tramway line and the breezy Nile-front location of the village seasonally transformed it into a summer night entertainment haven for the Egyptian masses. Dinning describes a "long line of *cafés-chantants* upon the shore, with their backs to the water." He adds that "they do not sleep; far from it. They are very wide awake. They are blazing with light and with the color of the tent-work roof and walls. There is an acre of small tables before the dais [stage] of each and beneath the lamps they are thronged." Dinning also observed that there were a significant number of Egyptian women sitting in these cafés, and that he and his companions were often the only non-Egyptians in attendance. The shows in the outdoor cafés of Rud al-Farag were similar to the ones taking place in the more elaborate venues in Imad al-Din Street and the greater Azbakiyya district, though the stars performing at Rud al-Farag were typically not as well known. The musicians on the stage sat in a *takht* arrangement similar to the scene in the Eldorado, and on the "flanks sit the instrumentalists." The majority of the primary performers and singers were also women, as Dinning elaborates:

> There is one lady in the center; between sits the chorus. The lady is the cynosure of all eyes. She may or she may not wear a yashmak. She may or may not have an instrument. One thing is certain; she is the soloist, and the only one there; and she is conscious of the importance of the position. If there is singing in chorus she leads it; but for the most part she sings stanzas alone, to which the men add a chorus.[73]

Unlike in the larger and glitzier Azbakiyya entertainment district, where there were always a significant number of European residents and tourists, Rud al Farag's theaters and casinos were predominantly patronized by Egyptians. During both World Wars, non-Egyptians may have been in the majority in Azbakiyya's cinemas, theaters, and casinos, as Cairo was inundated by allied troops thirsty for drink and entertainment. Even during the wars, however, non-Egyptians rarely visited Rud al Farag, which was a much cheaper and smaller entertainment district with one major open-air theater for plays, and about a dozen cafés and bars with stages mainly for summertime music concerts. Whereas the primary season for entertainment in the greater Azbakiyya district in downtown Cairo was during the cooler months from October to May, Rud al-Farag was known almost exclusively as an affordable summertime entertainment haven for the vast majority of Cairenes, who could not afford to travel to Alexandria during the hot summer months.[74] The nightly main events at Rud al-Farag were the plays that were performed in its large outdoor theater. As Fathi Radwan, the ultranationalist co-founder of Misr al-Fatah (the Young Egypt Party), recalls from his childhood in the second decade of the twentieth century:

> [My uncle] took me to the Nile coast of Rud al-Farag, which used to be a summer haven for the masses of Cairo. There, in the [summer] evenings, they can enjoy a nice breeze radiating from the Nile, while enjoying cold refreshments and secondhand imitation performances of some of the larger more successful theatrical troupes. Fawzi Munib for example performed the plays of Ali Kassar and Yusuf 'Iz al-Din and Fuad al-Gazayirly performed the plays of [Naguib] al-Rihani. So, what I ended up missing from the authentic plays of these famous theater troupes I watched in these cheaper more humble theatrical remakes.[75]

The semiprofessional theater troupes described by Radwan, and a few others, performed in Rud al-Farag in the summers but traveled continually throughout the Egyptian countryside, typically heading to Upper Egypt in the winter.[76] It is important to emphasize that before air conditioning became more common in Egypt in the 1940s and 1950s, indoor theaters and cinemas were simply not usable in the summer months, hence the necessity for outdoor entertainment venues. The Nile-front, open-air Rud al-Farag theater was described as being built "in the heart of a grove of eucalypti," with vines partly growing on its stage.[77] To be sure, however, the most important attribute of this stage to the thousands who attended these plays was the pleasant Nile breezes, which made the hot Cairene summer nights much more tolerable.

Night Lights, Cabarets, and Neon Signs

> What we think of as night life includes this nocturnal round of business, pleasure and illumination. It derives its own, special atmosphere from the light that falls onto the pavements and

> streets from shops (especially from selling luxury goods), cafés and restaurants, light that is intended to attract passersby and potential customers. It is advertising light—commercialized festive illumination—in contrast to street light, the lighting of a policed order.
>
> Wolfgang Schivelbusch, *Disenchanted Night*[78]

Electric lights were essential for a regular urban nightlife and were especially important for the indoor and outdoor illumination of theaters, casinos, and cafés. Before electricity was prevalent, theaters effectively made use of municipal gas lighting, though gas lighting had distinct disadvantages for indoor use, especially in crowded theaters. Unlike gaslight, electric light produced by incandescent lightbulbs does not give out noxious fumes, including carbon monoxide and carbon dioxide, and leaves no ammonia or sulfuric acid residue, which can damage fabrics and paint. It also does not raise indoor air temperature, which can be unbearable on Egyptian summer nights.[79]

Outside the theaters, electric lighting was arguably even more useful for theater companies and advertisers. Large and powerful arc lights, which were unsuitable for indoor use, were available as early as the 1880s, though they were improved and perfected in the first two decades of the twentieth century. Until the 1940s, these carbon arc lamps had a distinct advantage for outdoor use. They were very bright and capable of lighting large boulevards and squares, and for the candlepower produced, they were considerably cheaper than gaslight. A typical arc light produced "thousands of candle powers," and by the time of World War I, they could be focused like a floodlight, lighting up targets "up to six kilometers" away.[80] However, unlike incandescent lightbulbs, they produced carbon dioxide, created a flickering buzzing sound, and were a fire hazard, especially if they were used indoors. As early as 1912, Cairo had ninety of these powerful outdoor electric arc lamps, mostly positioned in Ataba al-Khadra Square and the other "principal places and squares" throughout the city, including of course Azbakiyya and the theater district. Eighty of those light fixtures were provided by the Cairo tramway company and ten by the Lebon Company.[81] The electricity for the ninety arc lamps in Cairo was provided gratis by the two companies. These figures do not include the many arc lights installed in Heliopolis (a suburb of Cairo) by the Heliopolis tramway company.

To be sure, providing electricity free of charge for public arc lights was not a random act of benevolent corporate generosity. Both the electric utilities and the tramway companies had to deal with balancing their electricity load. As David Nye has shown in his study of electricity in early-twentieth-century America, balancing the electrical load was critical, especially in the early days of electricity. To remain profitable, but also to keep their transformers and lines from shorting or overloading, the companies had to achieve balanced power consumption. The best ways to do this were to use up extra power though these arc lamps and/or to create consumer demand during off-peak hours.[82] In fact, lighting up streets and city squares near cinemas, theaters, and shopping districts incentivizes all of those theater, café, and business owners to keep their establishments open and, in turn, consume more power from these companies. Modern amusement parks like Luna Park, which opened in Heliopolis in 1911, was a perfect example of this strategy. The park's electrically operated rides and bright lights created yet another nighttime attraction, using the extra power generated during off-peak hours while also keeping the tramways and electric rails from Cairo to Heliopolis busy at night. The money made from ticket receipts for entrance to the park and for the various rides was simply icing on the cake.[83] Meanwhile, because of the expanding entertainment venues, the tramway companies now had thousands of nighttime riders using their trams to get to and enjoy the newly created nightlife. As early as the 1910s, as witnessed by the Australian officer Hector Dinning, "none of the restaurants seemed ever to close, and long after the last electric car had left the city loaded precariously to the trolley pole, the brilliantly lit cafés disgorged those who could afford a taxi to camp."[84] Trams in Cairo were supposed to stop running after 1:00 a.m., but many entertainment venues continued later into the morning.

From the 1920s to the 1940s, the well-lit theater and now cinema district at Imad al-Din Street would draw thousands of nighttime theater and cabaret customers, who would mostly arrive by tram, though increasingly by bus as well. Colorful neon lights were used in Egypt as early as the 1920s. Back then, neon was sparingly used to light up the signs of the major theaters, cabarets, cinemas, and a few of the large department stores.[85] By the 1940s, there were several Egyptian firms manufacturing and installing fluorescent and neon lighting, making it much more affordable,[86] and neon signs were prevalently on display in most of the large squares in Cairo and Alexandria, and not just in the entertainment districts.

One of the most impressive neon-lit entertainment complexes of that era was Badiʻa Masabni's Casino Opera, located in the heart of Opera Square (see Figure 4.6). Badiʻa Masabni was born in Beirut to Syrio-Lebanese parents, and raised in Buenos Aries. In the early 1920s, she moved to Egypt, married Naguib al-Rihani in 1924, and by 1926, she had become the most influential entertainer and casino owner in Imad al-Din Street. She is credited by many with being the biggest patron and mentor of modern belly dance, and she helped discover and train some of Egypt's most successful belly dancers, including Tahiyya Karioka and Samia Gamal.[87] Masabni's Casino Opera was the last and most ambitious of her casinos, after she had owned and operated some of the most successful casinos in Imad al-Din Street and elsewhere in Cairo.[88] This large, air-conditioned entertainment complex was remarkably successful when it opened in February of 1945.[89] The enormous building included a rooftop dinning terrace, a bar, a restaurant, a cinema, and nightclub with a large circular stage.[90] By the 1960s, cinemas would replace many of the cabarets and casinos in Imad al-Din Street, and most of the dancing halls would migrate to Pyramid Street (Shariʻ al-Haram) in Giza (see Tables 4.3 and 4.4).

FIGURE 4.6. Neon Signs and Cairo Lights (ca. 1950).
SOURCE: Egyptian postcard photograph, Cairo by night, Opera Square.

TABLE 4.3. Cairo's Major Cabarets and Theaters (ca. 1950).

CABARETS AND DANCING HALLS	LOCATION
Auberge des Pyramides	Pyramid Street
Auberge du Turf	Imad al-Din Street
Casino Coubana	Mustafa Kamil Square
Casino Hilmiya Palace	Al-Hilmiya
Casino Opera [Badi'a Masabni]	Ibrahim Pasha Square [Opera Square]
Casino Tabaran [Kit Kat]	Imbaba
Casino Shahrazad	Alfi Bey Street
Casino Sofar	Al-Ahram Street

THEATERS	LOCATIONS
Andalus Garden Theater	Qasr al-Nil Bridge
Azbakiyya Theater	Al-Ataba al-Khadra Square
Lycée Français Theater	Yusif al-Gindi Street
Opera House	Ibrahim Pasha Square [Opera Square]
Opera Malak Theater	Imad al-Din Street
Ritz Theater (Naguib al-Rihani)	Imad al-Din Street

SOURCE: Al-Qahira: *Kharitah Siyahiyya: al-Amakin al-Hamah wa Khutut al-Tram wa al-Utubis* (Cairo: Maslahit al-Siyaha, 1953).

Movie Theaters

> The technology of the talkies both culminated and celebrated the modern soundscape in ways that were impossible to ignore, for sound motion pictures gave voice, not just to the silent shadows on the silver screen, but to modernity itself.
>
> Emily Thompson[91]

The availability of electricity and the rapid innovation in lighting technology would soon come together to produce cinematographic advancements, allowing for cinemas, which provided yet another "modern" social space for entertainment and socialization (see Figure 4.7). The earliest recorded showing of short silent films in Egypt was in Alexandria on the night of November 5, 1896. Short Lumière Brothers' productions were shown daily every half hour from 5:00 p.m. to 11:00 p.m.[92] By early 1897, there were already dedicated movie theaters operated by the French company Cinématographe Lumière in Alexandria and

TABLE 4.4. Cairo's Major Movie Theaters (ca. 1950).

MOVIE THEATERS	LOCATION
Cairo Palace	Galal Street
Cosmo	Imad al-Din Street
Diana Palace	Alfi Bey Street
Al-Huriya	Heliopolis
Al-Nasr	Ibrahim Pasha Street
Azbakiyya Garden	Al-Ataba al-Khadra Square
Femina	Imad al-Din Street
Heliopolis Palace	Ahram Street (Heliopolis)
Le Pigalle	Imad al-Din Street
Lido	Imad al-Din Street
Lux	Imad al-Din Street
Metro	Sulayman Pasha Street
Metropole	Fuad I Street (Midan Halim)
Miami	Sulayman Pasha Street
Odeon	Al-Nimr Street
Opera	Ibrahim Pasha Square (Opera Square)
Radio	Sulayman Pasha Street
Rivoli	Fuad I Street
Roxy Palace	Heliopolis
Royal	Ibrahim Pasha Street

SOURCE: Al-Qahira: *Kharitah Siyahiyya: al-Amakin al-Hamah wa Khutut al-Tram wa al-Utubis* (Cairo: Maslahit al-Siyaha, 1953).

Cairo, showing what were mostly twenty-minute silent films.[93] Until "proper" cinemas became abundant and institutionalized as the place to watch movies, movie projectors were used in a variety of public spaces. As I mentioned earlier, by the turn of the twentieth century, many of the *cafés chantants* and cabarets used short silent films to entertain their audiences between music and dance routines. By the 1910s, silent films were popular in Egypt among all classes, and this trend continued until the "talkies" took over in the 1930s.[94] As a 1912 visitor to Egypt summed up, "the cinematograph entertainment is everywhere in evidence, good, bad and indifferent."[95] In the hot summer months, open-air cinemas, in particular, were quite popular. Describing Heliopolis near the end of World War I, Martin Briggs observes that the "most striking attractions of the place, the open-air cinemas, still remain. One may sit on the terrace of a hotel or a café for an hour or two any evening, watching 'the pictures' unfolding their blood-curdling dramas, for the price of a cup of coffee or a drink."[96]

There is ample evidence that silent films were very popular among a wide section of Egyptians during the first third of the twentieth century. By the 1920s, movie production and exportation were an important component of U.S. trade and the United States Department of Commerce was pestering U.S. consular agents and American trade commissioners abroad asking for detailed reports concerning the popularity of American films abroad. In just such a report dated March 18, 1926, Raymond H. Geist, the acting American consul in Alexandria, wrote that "many American films are constantly exhibited, and, as I find, are popularly received by the public, particularly by the lower class Egyptians patronizing the cheaper theatres."[97] The Josy Film Agency, which was the largest American film distribution agency in 1920s Cairo, estimates that in 1926, American films made up 80% of all films shown in Cairo, as "the first rate Cinemas run nearly 70% and popular Cinemas about 90% of American films; French and other productions are used on average of 10%."[98] Although it may seem counterintuitive today, silent films were never "silent" as there was always a source of live or

FIGURE 4.7. Movie Theater and Café on Imad al-Din Street (1925). *The Splendid Crime*, starring Bebe Daniels, is playing at the theater.
SOURCE: Postcard photograph titled "Caire—Emad El Din Street."

recorded music that accompanied the movie track. Remembering the silent movies he watched in his childhood in the late 1910s and early 1920s, Fathi Radwan recalls a blind piano player who regularly played music in the Olympia cinema in Cairo.[99] By the early 1930s, however, "talkies" would begin to proliferate.[100]

Talkies: Indoors and Out

The same loudspeaker and vacuum tube amplifiers developed for radios and public address systems were adapted and synchronized for motion picture use by the late 1920s, dramatically changing public entertainment.[101] In Egypt, the first sounded movie theatre was installed in Cairo in August 1929.[102] A British economic report found that by the early 1930s, "sound films are now in universal demand." The report indicated that in Cairo, as early as 1932, "owing to the advent of cheaper talking equipment, almost every Cinema, is now equipped with some sort of sound installation."[103]

In the mid-1930s, there was a lot of competition among the cinemas in Cairo and Alexandria, which prompted many of the movie theaters to compete for acquiring more popular films (see Figure 4.8). Because of recent taxations and tariffs imposed by the Egyptian government, however, this left very little profit margin for many of the local movie theaters. This problem was compounded by the fact that without air conditioning, indoor cinemas could not profit and attract a large audience during the hot summer months. Open-air garden cinemas in Cairo, Alexandria, and elsewhere in Egypt, which operated between May and the end of October, were very popular in the 1930s. Their profit margins were also better, since they could seat more spectators, relied extensively on selling cold drinks to augment their profit margins, and "the films shown were usually less up-to-date and consequently cost less."[104] In the 1930s, only a few government buildings, including Egypt's parliament houses, were air-conditioned.[105] But after World War II, as the cost of both electricity and air-conditioning units began to decrease, more public and private buildings installed air-conditioning units. Theaters and cinemas in particular realized the importance of air conditioning to their bottom line.[106] As air-conditioned movie theaters became more prevalent in the late 1940s, they would gradually undercut the open-air theaters.[107] At that time, cinemas and theaters made sure to emphasize that their establishments were air-conditioned, in order to attract larger audiences during the summer months.[108]

Hollywood hegemony would continue in the 1930s and 1940s, as American films would remain popular, and representatives from most of the important

The Soundscapes of Modernity   151

FIGURE 4.8. The Strand Cinema in Raml Station Square in Alexandria (ca. 1939). The two advertised films are Heidi, starring Shirley Temple (1937), and *Topper Takes a Trip*, starring Constance Bennett and Roland Young (1939).
SOURCE: The author's collection of "tourist photos."

Hollywood studios had offices in Egypt.[109] However, the Egyptian movie industry would soon compete with American films in the local and, later on, even the regional market. The first two Egyptian sounded films appeared in 1932, achieving "a considerable amount of success."[110] With the establishment of Studio Misr in 1934, the Egyptian movie industry would produce movies that were popular and, more importantly, accessible to millions of Egyptians.[111] A 1937 British economic report described these early Egyptian sounded movies as "extremely popular with the numerous Arabic-speaking public and have been known to run for as long as two months, in comparison with which first class imported European or American films are rarely shown for more than one week at a time."[112] By the late 1940s, the Egyptian movie industry improved the quantity and quality of its films, and inside Egypt, those films were competitive with American films, maintaining "long runs when shown in the principal cities and towns." By that time, Egyptian films were also being exported to "neighboring Arabic-speaking countries."[113] This corroborates Emily Thompson's analysis of non-western films in India and elsewhere, where talkies in the local vernacular provided a path of credible competition with Hollywood. Thompson adds that "talkie technology increasingly served the nationalistic agendas of numerous countries and colonies around the world over the course of the 1930s and beyond."[114] This was certainly the case in Egypt, where popular Egyptian movies with believable everyday characters speaking in the Egyptian vernacular had a national and even regional cultural impact.[115] Beyond the cultural impact that the content of Egyptian movies had on audiences, the outdoor and indoor cinema spaces provided yet another loud "modern" space for mass socialization. The social, class, and gender implications of these physically embodied interactions have yet to be studied by historians of the modern Middle East.

Conclusion: The Sonic Impact of Electricity

> Between 1880 and 1920 electricity began to permeate modern, urban life. Local traffic systems, lifts, the telephone, radio and cinema as well as a constantly growing number of household appliances would have been inconceivable without electricity.
> Wolfgang Schivelbusch, *Disenchanted Night*[116]

The introduction of electricity in Egypt was an entirely uneven process that began at the turn of the twentieth century in the wealthy and modern urban areas and would slowly expand to middle-class and working-class areas from the 1920s to the 1950s and beyond. Nighttime visitors to Cairo during the first decade of the twentieth century often noted the stark contrast between well-lit areas of the city like the Azbakiyya district, and the almost complete darkness of old Cairo, including the touristy al-Husayn district. By the 1920s and 1930s, more middle-class neighborhoods were lit up by electricity, though in most cases there was probably a lag, with electric streetlights and the infrastructure for electricity reaching a neighborhood first, and later on those that could afford it connected electric service to their private residences. By the 1950s, many working-class neighborhoods in the major cities had access to electricity, and over time, electric appliances, especially radios, became cheaper and more households had access to them. Certainly by mid-century, the buzz of electric fans and refrigerators and the sounds of blaring radio sets characterized many professional middle-class households. By the 1960s, Nasser's state-owned factories were producing transistor radios and other electric appliances that were affordable to many working-class families. To be sure, many Egyptian villages remained beyond the reach of electrification and even of running water well into the second half of the twentieth century, though by the end of the century, the process was virtually complete.

As I examined in this chapter, the sonic implication of electricity was enormous, as it not only allowed for radios, cinemas, loudspeakers, and tramways, but just as importantly, it resulted in the proliferation of electric lights that forever changed the sounds of the Egyptian night. Indeed, just as elsewhere, municipal gaslight and electric light created, for the first time, a regular nightlife, with all of its entertainment, leisure, commercial, and sonic implications. Wherever electric lights and tramway lines penetrated, night-time socialization and entertainment followed. Tramways in Egypt ran daily until 1:00 a.m. and played a pivotal role in creating a secure, well-lit nighttime transportation network feeding Cairo's and Alexandria's nightlife. Just fifty years earlier, before the tramways and before the street were lit with gas and electric lights, people did not regularly travel beyond their immediate neighborhoods at night. But having well-lit streets, coupled with effective, cheap, and regular transportation provided an entirely new, vibrantly loud nighttime economy.

Electricity made radio, films, and movie theaters possible and dramatically enhanced the sound quality of records, making them more popular and pleasing to the ear. The importance of electrical amplification for listening to everything from radio to records and for simultaneous and effective public communication to thousands of listeners during large public events, speeches, and concerts should not be underestimated. In one way or another, electricity and all of its associated lights, instruments, microphones, speakers, vehicles, and gadgets had a tremendous impact on every aspect of twentieth-century Egypt, including its sounded environment.[117] In the mid-1940s, microphones and sound amplification even changed what was arguably the most identifiable keynote sound of urban Egypt, when some of the major congregational mosques started using loudspeakers to deliver the call to prayer.

The Sounds of Public Spectacles

Between Ordinary People and State Legitimacy

> Those who wail or shriek loudly during a funeral [or wake], to the point of disturbing nearby residents, will be charged a fine ranging from ten to thirty piasters and may be imprisoned from one to five days.
>
> Clause 346 of the 1893 Legal Sentencing Guideline[1]

> His Majesty [King Fuad] was followed by Zaghloul Pasha [Saad Zaghlul] and the Ministers, and they took their seats on the dais to the left of the throne. . . . [The] Grand Chamberlain then advanced and handed the King his speech, which in his turn, his Majesty handed Zaghloul Pasha who read it aloud. . . . One hundred and one guns were fired while the speech was being read.
>
> *The Egyptian Gazette*, March 17, 1924[2]

5 The Sounds of Weddings and Funerals
From Brass Bands to Wails and Ululations

AT THE TURN OF THE twentieth century, traditional Egyptian weddings and funerals involved multiple ceremonial phases and a variety of public auditory and visual displays. There were social pressures and a nagging obligation to announce widely both weddings and funerals, and elaborate street processions were integral to both. The entire neighborhood, expanded well beyond the limits of a family's typical social circle, had to be invited to publicly morn or celebrate, and attendance, especially for funerals, was socially mandatory. During traditional weddings, maximizing the sounds of celebration was essential for declaring to all within sonic range of the festivities not only the joy of the wedding but also the legitimacy and the intended permanence of the new coupling. Although weddings were naturally much louder, with more varied sounds, funerals had their own loud sounds, mixed with moments of somber silence. Of course, the gradual utilization of loudspeakers, which started in the 1940s, exponentially increased the sonic range of these events.[1]

This chapter examines the sounds and sights of traditional Egyptian weddings and funerals during the first half of the twentieth century. The changing roles of street music, singing, loud funerary grieving, and other verbal and nonverbal vocalizations will especially be analyzed. Throughout this examination of wedding and funerary soundscapes, I will especially focus on the street and public manifestations of these everyday events. First, I examine some unique and important keynote vocalizations of traditional weddings and funerals. Then I briefly survey some of the legal implications of the raising of the temporary, large, pavilion-style canopy tents that were (and still are) used to extend the space of mourning or celebration into Egypt's streets and alleyways. I then examine popular street music, including the marching brass bands that were used regularly in Egypt for all kinds of public occasions. The rest of the chapter delves into the varying and changing sounds of funerary and wedding processions and rituals. The chapter concludes with a discussion of the aforementioned attacks against many of these traditional rituals by both Islamists and secular modernists. This dual assault on many of these sounded traditions took a toll, as many of them died out or steadily declined during the second half of the twentieth century.[2]

As I have emphasized throughout, the constant attempts at silencing and rendering invisible itinerants in the Egyptian street were in large part motivated by class. Class anxiety and fear of the crowd also played a role in the ongoing attempts at marginalizing, controlling, and in many cases limiting large group celebrations. These class apprehensions manifested themselves especially in regard to popular public expressions ranging from traditional and loud celebrations of weddings, funerals, and circumcisions to varying popular religious celebrations of *mawalid* (saint's day festivals), and *zars* (exorcisms). Most of the attacks on these popular traditions and rituals invoked a modernist critique of the musical, dancing, and general embodied nature of these events.[3] The one point that all the "modernizing" camps agreed on was their belief in the general ignorance of the vast majority of the population and the urgent need for education, reform, and uplift. Class distinction was especially important for Egypt's growing middle classes as they self-consciously attempted to define and separate themselves from most ordinary Egyptians. More specifically, this chapter examines these class distinctions through analyzing the changing public displays and acoustics of Egyptian wedding and funerary processions. Casting the fellahin and the Egyptian working classes as raucously loud, vulgar, and unrefined, allowed the growing middle classes to carve out for themselves the role of cultured, pious, and sensible citizens, albeit with the responsibility (theoretically at least) to educate, uplift, and consequently, quiet the masses.

Egyptian Women and Loud Nonverbal Vocalizations

> If geography was one determinant of the syntax of sound, social class was another. . . . Among educated listeners in early modern England there was a pronounced distrust of nonverbal sounds. The whoops and hollers of countryfolk and lower-class craftsman might be amusing in a pageant or a masque but such sounds marked the boundary between civility and barbarity. . . . The Irish "hubbub" in times of mourning, argument, or celebration struck horror and contempt in English ears.
>
> Bruce Smith, "Listening to the Wild Blue Yonder"[4]

Loud verbal and nonverbal vocalizations, broadcast individually or in unison, played a big role in modern Egyptian mourning and celebratory rituals. The first news of a death in a traditional Egyptian neighborhood was almost always the loud screams and shrieks of female family members. Similarly, the piercingly loud sounds of *zagharit* (sg. *zaghruta*), or loud ululations, are a clear signal that a wedding or another happy event is about to take place. In Egypt, both loud wailing or screaming upon the death of a family member and ululating to celebrate important life events are unequivocally gendered performances, almost exclusively sounded out by Egyptian women. This certainly contradicts and stands out from the traditional daily dynamics of gender modesty, yet in times of mourning and celebration, loud public vocalizations from most Egyptian women are widely expected and accepted.

Zagharit are among the quintessential Egyptian celebratory sounds, heard especially on or before weddings. A single *zaghruta* can last from a few seconds to almost a minute. Its sound may vary, but typical Egyptian *zagharit* are high-pitched howls, emanating from deep in the throat, with a wavering trilling effect created by moving the tongue rapidly back and forth with the mouth wide open.[5] Egyptian women usually lift one of their hands above their mouths as they are ululating to direct and project the sound further. As a manifestation of unadulterated joy, the sound of *zagharit* projects a sense of happiness beyond what verbal expression can convey, and doubles as a calling card of sorts and an invitation to the rest of the neighborhood to share in this joy. Depending on the occasion, *zagharit* can be accompanied by celebratory dancing, especially if performed in a setting with live music. Beyond weddings, *zagharit* were also almost obligatory during circumcision processions, birth celebrations, graduations, pilgrimages, releases from prison, and any other celebratory events.[6]

On the other side of the emotional spectrum, loud screaming and wailing was expected upon news of a death. Unlike *zagharit*, which are rarely accompanied with any words, wailing is often mixed with short, two- to three-word, verbal declarations of grief. There is an embodied aspect to wailing, as those expressing such uncontrollable anguish could tear their clothes, beat on their chests, and more commonly, slap their own faces with their open hands. As described later, there was also a long tradition of hiring professional female mourners to scream, wail, and lament during funerary processions. Although wailing and shrieking aloud were performed mostly for deaths and funerals, they could also be heard during other sad occasions. For example, mothers, daughters, and sisters might sonically mourn arrested, imprisoned, or even conscripted family members.[7]

The purpose of both wailing and ululations is twofold: quickly announcing the actual event to the neighborhood, and beginning the process of sharing the joy or grief with family, friends, and neighbors. Even when performed indoors, wailing and ululations were meant to project outward so others in the neighborhood could share in the sad or happy occasion. If the women happened to be indoors, the ululations or wails were often projected through an open window or balcony door for maximum public dissemination. There is also a dialogical component to both ululations and wailing, as the rest of the women in the community were expected to immediately respond in kind, sonically propagating the noise well beyond the immediate neighborhood. These vocalization dialogues continued in varying ways throughout the celebratory or grieving periods, creating a sonic emotional connection throughout the community. Despite the widespread performances of both ululations and wailing, since the middle of the twentieth century, both of these celebratory and grieving vocalizations have declined. With increased urbanization and suburbanization, the intimately close neighborhood structures changed and so did communal relations, especially among the growing middle classes. Other forms of communication gradually complemented and for many supplanted the more traditional vocal methods, which were increasingly viewed by many as either backward or vulgar. Here, yet again, class and class formation played a key role. Upper-class and upper-middle-class urban Egyptian women rarely if ever wail or ululate. Remaining publicly silent on occasions of celebration or mourning, especially for an upwardly mobile, middle-class woman, gives a clear signal of class distinction, differentiating her and her family from the urban working class and the fellahin.[8]

Public Processions and Building *Shawadir* in the Streets

> Passing through the Abbasieh [Abbasiyya] quarter, we always came, sooner or later, upon a wedding. The different stages of a native marriage require, indeed, so many days for their accomplishment that nuptial festivities are a permanent institution in Cairo. . . . One day, upon turning into a narrow street, we discovered that a long portion of it had been roofed over with red cloth; from the center of this awning four large chandeliers were suspended by cords. . . . In the roadway, placed against

the walls of the houses on each side, were rows of wooden
seats; one of these rows of seats was occupied by the band,
which kept up a constant piping and droning.

Constance Fenimore Woolson,
Mentone, Cairo and Corfu (1896)[9]

Most of the secular and religious celebrations and the funerary commemorations that I discuss in this chapter took place at least in part, in the public streets. This was mainly done to accommodate larger numbers of people, but it also served as a sort of loud billboard, displaying and announcing the event for all to see and hear. The need to propagate the news and celebrate these events with the entire neighborhood necessitated finding public venues large enough for all to participate. This was especially true for weddings and funerals as both had attached to them tremendous social obligations for active participation by the surrounding community.

In traditional urban areas, if the families could afford it, both weddings and funerals involved elaborate processions. Funerary processions, though somber, could be quite elaborate and involved a variety of sonic displays. *Zaffahs*, or celebratory processions, for weddings and other happy occasions always involved loud music, singing, and sometimes dancing. Circumcision celebrations and most neighborhood receptions for pilgrims returning from the haj tended to be smaller, but also involved public processions. Basic public processions to the cemeteries and *zaffah* processions for weddings and other events were affordable for most families, as the size and elaborations of these events was relative to the costs allotted to them. It was common, for example, for families to save costs by doing a joint circumcision celebration for several of the neighborhood boys or grouping a family wedding with circumcisions. Cost sharing was also common for haj processions, celebrating the sending off or return of neighborhood men and women to or from a pilgrimage to Mecca.

One characteristic landmark of a large wedding or a funeral was a temporary large and square canopy-tent structure, often with dark-blue, red, and white arabesque geometric patterns. These colorful *shawadir* (sg. *shadir*) were frequently erected in side alleys and streets in order to accommodate all the visitors and invitees. *Shawadir* were not just for weddings and funerals; they were also used for political campaigns and rallies, *mawalid*, Eid celebrations, and other large public events that necessitated a temporary public shelter and a visible street presence. They were always well lit and decorated, though they ranged in size and opulence based on the financial capabilities of the event sponsors. As soon as electricity became available at the turn of the twentieth century, many of the *shawadir* featured electric lights.[10] By the 1940s, many were also equipped with microphones and speaker systems, which dramatically increased the decibel level of the music, speeches, and in the case of funerals, Quran recitations, which were broadcast to the entire neighborhood. For families who could not afford the raising and decorating of a *shadir* for a funeral or a wedding, the main location for the event would be their home, though often rudimentary reed mats and chairs were placed outside on the sidewalks or streets, to accommodate more people.

The Egyptian government was not overly concerned with smaller wedding and funerary processions, as they did not obstruct traffic too much, but some laws existed that limited the extent of *shawadir* on public sidewalks, streets, and alleyways, and these large structures required official government permits. Some of the same laws discussed in Chapter 2 in regard to street side seating for cafés had clauses specific to these *shawadir* and to any other temporary street installations. It was in 1896 that the Egyptian Ministry of the Interior and the Public Works Ministry had amended a previously established 1885 law concerning the private use of public roads, making it more flexible and more accommodating, as it then allowed the temporary use of a portion of the public roads for weddings, funerals, and other communal commemorations and celebrations.[11]

This law provided an avenue for individuals to apply for and receive licenses, allowing them to temporarily use one-third to one-half of a public road to set up *shawadir*, lights, and other decorative installations for weddings, funeral, charities, *mawalid*, and the like. The charging of a daily fee, based on the number of square meters of the road or sidewalk that were in use, applied only in the case of weddings. The government perhaps viewed the application of a similar fee for funerary *shawadir* as being in poor taste. This wedding fee was decreased by 50% if the road used was unpaved or just macadamized. In Cairo, the authorities did not allow the use of certain busy roads for these purposes, and the emended 1896 law specifically forbade the use of the often crowded Muski, al-Nahasin, al-Sukariyyah, and al-Ghuriyah Streets. The key stipulation was that these installations could not obstruct traffic and must use only a part of the road. Finally, the law required that the roads and alleyways had to be returned to their original state after the temporary structure was taken down. All damages to the roads or sidewalks incurred due to the raising of *shawadir* or any other temporary structures had to be repaired and paid for by the event licensees.[12] At first, this law was instituted in Cairo and Alexandria, but by the turn of the century, it was applied in most of the cities and towns in Egypt. Of course, as with the related laws concerning alfresco cafés and bars that I discussed in Chapter 4, enforcement was irregular, and there were always inventive ways to thwart actual enforcement, especially if the event were being held in a small alleyway or street.

Popular Street Music and Hasaballah Brass Bands

> At the end of the street, a well off family celebrated its eldest daughter's marriage to a respected government employee with a large wedding in its large house overlooking the [Muqattam] Hills. I was invited along with my family to attend this wedding and though I don't remember many specifics related to the house itself, I do remember clearly that I sat close to the police band. For at that time they were employed to play at weddings for a small fee. These types of music bands spread all over the street a sense of glorious happiness and joy not just because of their music, but also because of the dignified uniforms they wore and their beautiful and brilliantly shiny brass instruments.
>
> Fathi Radwan, *Khati al-Ataba*[13]

Fathi Radwan's childhood memory of a police brass band playing at a wedding in the Sayyida Zaynab district of Cairo in the late 1910s is representative of an ongoing change in the sounds of Egyptian street music. By the turn of the twentieth century, if not sooner, brass bands had become synonymous with popular Egyptian urban weddings and celebrations, though by mid-century, loudspeakers projecting live and, later on, recorded music would predominate in most celebrations. The popularity of brass bands had begun to take root almost a century earlier, with the military reforms of Muhammad Ali (r. 1805–1848). Starting in the mid-1820s, Egyptian military music schools trained hundreds of musicians in western style marching music.[14] As I detail in the next chapter, these early military-trained Egyptian brass bands played mostly western music: various anthems, including the "Marseillaise," and dozens of other pieces of military-style marching music for the Egyptian army and for the wider public, especially during official parades and processions (see Figure 5.1).[15] As Adam Mestyan has shown, weddings for the Muhammad Ali family and possibly other elites in Egypt had employed military bands since at least the mid-1840s.[16]

FIGURE 5.1. Egyptian Army Band Returning to Barracks after the Departure of Cromer in 1907.
SOURCE: Egyptian postcard photograph, 1907.

Here though, we need to differentiate between the European martial music these bands played when they operated in their official capacity as either military or police bands, and the music they played for private events and celebrations. More importantly, many of the conductors and musicians of the Egyptian military bands later took part in the creation of dozens of commercial marching bands, playing mostly Egyptian tunes adapted to western music and instruments. Indeed, these syncretic Egyptian street brass bands created their own genre of music, playing a sort of fusion of traditional Egyptian music and tunes with western martial music and using mainly European instruments. Despite the western origins of their music, these street bands would soon be considered part of the baldi or *sha'bi* (working class) music tradition, and from the late nineteenth century onward, they were used in local celebrations, playing their own new syncretic sounds for all Egyptians to enjoy.[17]

A comprehensive early history of these offshoot commercial brass marching bands has yet to be written, but most likely, former members of the Egyptian military bands started them in their spare time or after they retired from their positions.[18] Some of these commercial brass bands could have been formed as early as the 1840s, when tens of thousands of Muhammad Ali's troops were put out of work as a result of the Convention of London of 1840 and the forced reduction of the Egyptian military from over 130,000 troops to just 18,000.[19] The availability of many hundreds of out-of-work trained musicians, and just as importantly, the abundant availability of brass instruments, bass drums, and cymbals provided the opportunity and means for enterprising former military band members to form their own groups. No doubt, the same thing happened post-1882, after Urabi's defeat and the downsizing yet again of the Egyptian military under the British occupation. In any case, certainly by the 1880s, there is ample evidence of these Egyptian urban street bands performing in regular, non-elite Egyptian weddings.[20] These private bands varied in quality and size and probably charged accordingly. There were certainly variations in the types of instruments used, as a surprised Alfred Cunningham found out on his 1911 visit to Cairo: "It is quite a common sight to see a native band, consisting of pipers and drummers, playing Scottish airs, and leading a wedding cavalcade. So great a hold has the much maligned bagpipe secured upon the affections and presumably upon the ears of the Egyptians that there are now in the bazaar quite a number of native shops devoted solely to the selling and repairing of bagpipes."[21]

The most famous of these commercial brass bands was the Hasaballah Band (Firqit Hasaballah), established sometime around the turn of the twentieth century by a certain Muhammad Ali Hasaballah, a former music conductor or musician in the Egyptian military. The band continued to operate in Cairo under the ownership of the Hasaballah family for several generations.[22] The headquarters of the band—along with those of many other brass bands, belly dance troupes, and other music *takhts* (ensembles) and musicians—were in Muhammad Ali Street, which was the center of music life in Cairo.[23] Sometime after the original Hasaballah Band moved to Muhammad Ali Street, it expanded its reputation as the premiere popular *baladi* (local Egyptian) brass band, and the name Hasaballah became synonymous with Egyptian-style marching band music as it entered the Egyptian vernacular. By the 1920s, all of the many Egyptian marching brass bands were being called Hasaballah bands, and the entire music genre was called Hasaballah music (*mazikit* Hasaballah).[24]

One of the reasons for the rapid popularity of Hasaballah marching bands was the importance of memorializing and celebratory processions, be they secular or religious, in Egyptian popular culture. Traditional Egyptian weddings at the turn of the twentieth century had as many as half a dozen *zaffahs* (celebratory processions). There were also celebratory processions for newly circumcised boys and for returning and departing pilgrims, and many other processions were associated with the dozens of *mawalid*. All of these events required music, and the mobility of these bands and the big sounds projecting from their brass instruments perfectly fit these occasions. No doubt, the impressive appearance of these bands, with their uniforms and shiny brass instruments, also played a role in their adoption and appreciation by the Egyptian masses.

Long before Hasaballah brass bands burst onto the scene, traditional bands and musicians had been active in the Egyptian streets for hundreds of years. These more traditional bands, composed of at most half a dozen musicians using traditional instruments, were more common and cheaper to hire than the brass bands. Typically, in these bands the drums predominated, such as the *darbuka* (goblet drum), *tabla baladi*, *riqq* (Egyptian tambourine), and *daf* (large frame drum), though Upper Egyptian sa'idi bands are characterized by their use of wind instruments like the *mizmar*. Employing a traditional Upper Egyptian band with its distinctive *mizmar*, a conical double-reed woodwind instrument, and drums was especially popular for wedding *zaffahs*. For most celebrations, however, if funds were tight, a couple of hired percussionists might do or, family members could use a borrowed drum or two to sound out and pace some of the celebratory songs.

In the late nineteenth century, the use of military and police brass bands in Egyptian *zaffahs* and celebrations was considered exclusive and prestigious, and was available primarily to well-connected elites. As private Hasaballah groups popularized this genre of music, and Egyptianized it starting in the early twentieth century, brass bands became desirable and available to the middle classes and, soon after, became extremely popular and accessible to the Egyptian urban masses. Predictably, as it lost its sense of exclusivity, this style of music was shunned by the growing middle classes and vulgarized as "low class."[25] The ever-present drive for class distinction can be a powerful force in setting the always-changing aesthetics of taste, musical or otherwise. Just like the celebratory *zagharit* and the wailing laments, which were once synonymous with the soundscapes of mourning and celebrations in Egypt, Hasaballah bands are in decline, and unlike the former, may not recover their brief albeit sonorous role in the Egyptian soundscape.

Another marker of class distinction was the increasing ability of middle-class families to rent indoor spaces in hotels and dance halls, which by the second half of the twentieth century were well equipped with loudspeakers. This move indoors largely confined the main part of the celebration to only the invited guests.

Weddings and the Sounds of Celebrations

> Sprinkle water on the street . . .
>
> The bride of our dearest is coming
>
> Sprinkle Coca-Cola on the street . . .
>
> The bride of our dearest is riding here
>
> Sprinkle perfume on the street . . .
>
> The bride of our dearest is beautiful
>
> Sprinkle the street with electricity [i.e., light] . . .
>
> The bride of our dearest is welcome here
>
> Yes, O dark neighborhood [*harrah*] . . .
>
> we are the one who will light it up
>
> Yes, O hush-hush neighborhood . . .
>
> We're the ones who will give you a loud voice[26]

"Sprinkle Water on the Street" (*Rushuh al-shari' maya*) is an old, improvisational Egyptian wedding song with multiple and varying lyrics. Before the asphalting of Egypt's roads (see Chapter 3), sprinkling water in the streets prevented dust and sand from blowing everywhere; so spraying the streets with water was especially a necessity before any large procession. By calling for the figurative spraying of the street with sweet drinks or fragrant perfume, the song highlights the importance of the special occasion. The reference to Coca-Cola means that this version was sung in 1948 or after, since that's when the soft-drink company started bottling its products in Egypt. In older versions of the song, *kazuzah* (fr. *gazeuse*, or soda) or *sharbat* (sherbet) is called for instead. In any case, this is one of the many songs traditionally sung in the bride's street procession (*zaffit al-'arusah*). The last two lines of the song, referring to loudly and openly celebrating the wedding, signal the importance of wide-open, unabashedly loud communal celebrations.

Bright lights, noisemaking, singing, and music are essential for declaring the marriage union to the entire neighborhood and beyond. This traditionally and openly legitimates the union of the newly established couple in the eyes of the community at large. Traditional Egyptian wedding celebrations typically lasted for at least three days, and up to a week if the family could afford it. As mentioned, for those with limited means, the base for the wedding celebration is set up in the family house, with chairs lined up in the streets to extend the capacity of the event.[27] Those with some means usually set up a well-lit *shadir* in the neighborhood to accommodate more guests, and to provide a space for the musical entertainment, which was then available for the entire neighborhood to enjoy. Wedding *shawadir* often had a stage for the singers, musicians, and dancers, and rented chairs for the guests, and depending on the neighborhood, by the late 1920s most were supplied with electric lights.[28] After World War II, many *shawadir* were also equipped with microphones and loudspeakers, which dramatically expanded the acoustic range of the music and the rest of the festivities.

The Sounds of Wedding Processions

> Before the wedding, the bride is conducted in gala attire and with great ceremony to the bath. This procession is called "Zeffet el-Hammam." It is headed by several musicians with hautbois and drums; these are followed by several married female friends and relatives of the bride in pairs, and after these come a number of young girls. The bride follows, under a silken canopy, open in front and carried on four long poles by four men. . . . In Cairo, however, this canopy is generally replaced by a carriage of some kind. . . . The procession moves very slowly, and another body of musicians brings up the rear. The shrieks of joy which women of the lower classes utter on the occurrence of any sensational event are called zagharit. The bride is afterwards conducted with the same formalities to the house of her husband.
>
> Karl Baedeker, *Egypt and the Sudan* (1914)[29]

The purpose of these *zaffahs* and festive musical marches was to methodically and deliberately parade around town while sonically and visually publicizing the happy event in order to draw the rest of the immediate community into the celebration. Wherever the *zaffah* went, there was music, hand clapping, dancing, and singing, almost always punctuated by *zagharit*. Bands of musicians, sometimes dancers, jugglers, and other entertainers were hired for these events, or if the budget was limited, family members and friends supplied the music and entertainment themselves. Among the characteristic sights of pre-twentieth-century wedding and pilgrimage *zaffahs* were decorated camels carrying large kettledrums and drummers (see Figure 5.2).[30] Most bands and musicians, though, traveled on foot, as the processions marched at a very slow pace and

FIGURE 5.2. Cairene Marriage Procession (1898).
SOURCE: New York Public Library, Catalog ID-B-number b10733511 (1898).

frequently stopped for dancing and merrymaking. This is one of the main reasons why marching brass bands could be so quickly and seamlessly introduced into the Egyptian street *zaffah* tradition.

To help pay for the hired bands and musicians, a tipping (*nuqta*) system was in play. As these processions inched forward to various points of interest around the neighborhood, merchants, storekeepers, and well-off neighbors and friends would halt the procession and demand a song or tune from the musicians, with tip money in hand. Typically, the bandleader loudly names the tipper and repeats after him or her, almost like a loud echo, the specific message or salute honoring the occasion.[31] The music then loudly continues after this formal verbal exchange. The general pattern for these processions was always the same, regardless of whether the *zaffah* was celebrating a wedding, a circumcision, returning pilgrims, or some other happy occasion.[32] Etiquette required making note of the people who were making these generous tips, as reciprocity was expected on future occasions. This functioned as an informal cost-sharing system that not only made these events financially possible but also built communal bonds.

Egyptian wedding *zaffahs* were more varied and multifaceted at the turn of the twentieth century than they are today. At that time, a traditional Egyptian wedding involved several of these celebratory musical processions over the course of at least a few days. As with circumcision and pilgrimage *zaffahs*, the elaborate wedding processions served, first, to maximize the exposure of the wedding ceremony to the entire community. In addition, each of these specific wedding *zaffahs* was designed to mark a particular threshold in the multiday wedding celebration. Before the wedding for instance, there was the bath procession (*zaffit al-hammam*), where the bride's close friends and female relatives celebrated, bathed, and pampered the bride to be. After the bath procession, there was the henna night (*laylit al-hinnah*), where the bride's hands and feet were decorated with intricate henna tattoos. The henna night was usually held the night before the wedding day.[33] Group singing and dancing were the order of the day, and female musicians often accompanied the family into the bath and were a part of the henna ceremony later on. For the groom, there was the barber procession (*zaffit al-mizayin*), which celebrated his pre-wedding grooming preparations.[34]

Douglas Sladen, who was visiting Cairo in 1910, describes a bridegroom's night *zaffah* he witnessed when he was returning at 1:30 a.m. from a dance at the Savoy Hotel. Despite Sladen's overtly judgmental and orientalist prose, his account is worth quoting at length, simply for his vivid description of the sounds he heard that night:

> First came a band on foot playing the usual mad tunes and making an awful noise over it (all the people in the hotel who had not been to dances and had not turned out to see what it was, were swearing about it next morning).... These were followed by other men carrying a sort of set piece—a large frame with about fifty lamps on it, arranged in four revolving circles. The bridegroom and his friends followed in a sort of circle, all facing this frame, and each of them carrying a lighted candle and a flower, and the procession wound up with more musicians. Every now and then, it stopped, and somebody sang something in the droning, Oriental way, which is more like reciting. The whole thing was exceedingly noisy and exceedingly picturesque.[35]

Typically, the largest and most important of the *zaffahs* was the bride procession (*zaffit al-'arusah*), which was held on the actual wedding day, when the bride was transported to the main venue (which might be her future house) for the primary wedding celebration[36] (see Figure 5.3). The music and the cel-

FIGURE 5.3. A Wedding Procession (ca. 1910).
SOURCE: Douglas Sladen, *Oriental Cairo*, 1911.

ebration would continue there through the early morning, until the bride and groom retired for the night. Perhaps the most unusual of the wedding processions was the *zaffit al-gihaz*, which roughly translates to the furniture and house accessories procession. As the name suggests, the entirety of the new groom and bride's household furniture and wedding trousseau was slowly paraded on a horse-drawn cart or truck for all to see. Just like the bride and groom in the other *zaffahs*, the furniture was slowly and deliberately serenaded with music, singing, and dancing throughout the streets. As I examine in the next section, traditionally, all the wedding *zaffahs* and many of the other wedding events have dozens of songs specifically addressing them. One of the songs for *zaffit al-gihaz* helps to explain that *zaffah*'s socioeconomic function by specifically and rhetorically asking during the procession:

> Where is your furniture oh bride? Let me see it!
> Where are the things your father bought? May God light his path![37]

Wealth displays were an important part of wedding celebrations, demonstrating the ability of the family of the bride and groom to provide for their sons and daughters. This applied as much to the contents of the trousseau and the *gihaz* as it did to the cost of the food and entertainment provided for the actual wedding.

Wedding Songs, 1900–1950

As described earlier, music and other sound projections were vital to traditional weddings and other mass celebrations in Egypt. Most towns in Egypt have professional musicians of varying skillsets to hire out for weddings and other occasions.[38] Those who could afford it set up a stage for the occasion and hired a full *takht* and professional singers and dancers to perform for a few days. Most hired a small band to perform and lead the musical celebration. Regardless of the economic means of the families holding the wedding, there were ways to share the expenses of the entertainment. Typically, just as with the *zaffah* bands, most of the wedding guests were obliged to tip the other wedding entertainers and musicians. In fact, the bride's and the groom's immediate families were not the only ones who displayed their generosity by providing for the wedding; many of the guests were expected to do so as well, especially those viewed as economically prominent within the community. Family, friends, and members of the community also gave money and other gifts to the newlywed couple. The expected worth of the gift was roughly determined by the wealth of the givers and their relationship to the wedding party. The name of the person giving money or a gift to the wedding party or the entertainers was loudly announced for all to hear, and reciprocity was the order of the day, as a comparable gift was expected when someone in the tipper's immediate family was married in the future.

Musically inclined friends and family members could also perform some of the singing in Egyptian weddings. This usually involved the use of a darbuka or some other drum and included communal hand clapping along with the singing. One singer (who may have had performing experience) led the singing, while friends and family members repeated the chorus.[39] There were regional differences in the types of instruments used and the songs and music played. Traditional Upper Egyptian music bands, for instance, relied heavily on the distinctive *mizmar* and the rebab (*rababah*), a bowed stringed instrument similar to a fiddle. Whereas Lower Egyptian delta music primarily used percussions and vocals. Aside from music, most Upper Egyptian and some countryside wedding celebrations are still characterized by the popping sounds of guns fired into the air. Unfortunately, local diversity is decreasing, owing to increasing cultural centralization and globalization, and the Egyptian middle and upper classes today often mix in western music along with Egyptian pop hits.

Complementing the music, were hundreds of elaborate traditional wedding songs, with dozens of specific, though always fluid lyrics designated for each part of the courtship and wedding process. Each of the *zaffahs* and wedding rituals, for instance, had its own selection of songs and lyrics. This vast anthology of wedding and marriage songs was geared toward a variety of topics, from providing marital advice for the bride to offering sexual advice for the wedding night (*laylat al-dukhla*), and even cautionary though humorous tales about controlling or conniving mothers-in-law. Most of these songs were not written down and had multiple, changing lyrics, with some of the lines improvised to include the names of the bride and groom or other information relevant to the specific occasion.

Many of the love and marriage songs have double meanings, with obviously "hidden" sexualized references. This is especially the case for most of the songs performed during the *laylat al-dukhla*, which literally means the night of consummation. These sexualized songs may take either the groom's or the bride's perspective:

> She took her shirt off . . . she put her shirt on . . .
> she went to her groom . . . at 1:00 a.m.
> He went up to tease her . . . went down to tease her . . .
> broke her bracelet . . . at 1:00 a.m.
> He went up with her . . . went down along with her . . .
> fell completely for her . . . at 1:00 a.m.[40]

Obviously, the line about the groom breaking the bride's bracelet refers to deflowering the young bride by breaking her hymen. The importance of the bride's virginity, real or perceived, is not as emphasized as it once was, but it is still important for a majority of Egyptians. Some of these sexualized songs, however, refer to more than just the physical act. For example, marriage and moving to a new household was quite dislocating:

> On the bed let's go . . . oh bride give me a kiss
> Oh groom my heart has fallen
> The lad with the handgun . . . his kisses are like mangos

> How lovely is cleaning [trimming] the okra ... Your mom and sister are cooking
>> While I am sleeping on your lap
>
> How lovely is cleaning [trimming] the zucchini ... Your mom and sister are cooking
>> And I am worthy of a kiss.[41]

Sung entirely from a young bride's perspective, this song is more than the sum of its parts. Aside from the obvious phallic and sexual symbolism, the references to the sister and mother-in-law cooking, while the new bride is "sleeping on [the groom's] lap," hints at some of the anxieties about relocating to a new household. Because the vast majority of brides had to move to a room in the groom's family's house or a house nearby, they had apprehensions about the shifting power dynamics within the new household. The following song excerpt depicts a classic example of a new bride testing out the subtleties of her new environment vis-à-vis her young husband and her new mother-in-law:

> Your mother feeds me *mish* [old cheese] ... And I don't like *mish*
> I swear I will sell the brass *tisht* [water basin] ...
>> and buy me some lamb meat my boy
>
> Your mother likes *kurat* [leeks] ... and I don't like *kurat*
> I swear I will sell a *qirat* [a piece of land] ... and buy me some lamb meat my boy[42]

This song, and many others like it, discuss the inevitable power struggle between the new bride and her new extended family. In the song, the new bride is testing her emotional capital over her new husband and flexing her economic muscles by indicating her willingness to sell her trousseau or even some land in order to have her way in running the household. The bride's main rival for both the affections of her new husband and her economic independence was usually personified in the figure of the mother-in-law. These exaggerated and always humorous depictions of the "villainous" and controlling mother-in-law are a staple of Egyptian humor, and are reflected in Egyptian jokes, proverbs and popular culture. Egyptian proverbs revealing anxiety about mothers-in-law are still currently in circulation, and they include, "I would rather have a flood or a fire rather than my mother-in-law in the house" (*al-mayah wa al-nar wala hamati fi al-dar*), and "a mother-in-law is like the flu" (*al-hama hummah*).[43]

The loud *zaffah* tradition, with its multiple musical street processions, and the singing and celebration that take place during the actual wedding were hallmarks of Egyptian marriage ceremonies. Maximizing the sounds of the celebration served the vital societal function of declaring to all within sonic range the legitimacy of this coupling and its intended permanence. Of course, the gradual deployment of loudspeakers exponentially increased the sonic range of weddings and other celebrations. However, in many ways, the diversity of the multiple traditional songs and the number of *zaffahs* have steadily decreased, especially among the middle and upper classes. This is in part due to the ongoing cultural and media centralization taking place throughout the twentieth century, with Egyptian State Radio and the record and movie industries creating new, more homogeneous musical tastes among Egyptian cultural consumers.

Since the 1950s, many middle-class and upper-class weddings have consolidated the many *zaffahs* to just two. The first is usually held at the wedding venue (often indoors in a hotel hall), where the bride and groom enter the wedding venue together along with a host of musicians, headed by a belly dancer, and playing a few couplets of only one or two traditional songs. The *zaffah* then proceeds slowly, making the rounds so everyone can see and participate, until the bride and groom eventually reach the *kosha*, two large throne-like seats elevated on a small platform. The other modern *zaffah* takes place at the end of the wedding, typically at 2:00 or 3:00 in the morning, when the newly married couple are driven to their new apartment. The "music" for this *zaffah* is provided by a parade of cars belonging to family and close friends who beep their horns rhythmically and collectively to honor the bride and groom. As with *zagharit*, other drivers who are not part of the wedding often beep their car horns in support, sharing in the couple's joy.

Sounding Out Deaths and Funerals

> It happened on the first night of our settling in Bab-el-Bahar [a suburb of Cairo] that "about midnight" there was a "great cry;" for in a neighboring house one was just dead. No one who has heard that sudden cry breaking the deep stillness of night can ever forget its thrilling effect. Then came the piteous wailing that seemed to speak of sorrow without hope: the mother of the family was taken, and the children's shrieks and sobs mingled with the plaintive cry of, "Oh, Aneeseh! Aneeseh!" from the sisters or friends who vainly called on her who could no longer answer them—who had no longer a name on earth.
>
> Mary Louisa Whately, *Child-Life in Egypt* (1866)[44]

In a traditional city quarter or village, the very first indications that someone has passed away are the almost primordial shrieking screams and wails of the deceased's female relatives alerting the entire neighborhood of the passing of a loved one.[45] Just as with the celebratory *zagharit*, female friends and neighbors echo the wails of the primary grievers in an almost dialogical conversation, sonically broadcasting the news even further afield. Sonically joining in with the grief of your friends and neighbors by responding to their wails in kind is a courtesy expected by all neighborhood women. This was of course more practical and functional in the years when most urban Egyptians lived in tightly packed, enclosed *hawari* (sg. *hara*) or city alleyways, that is, up to the middle of the twentieth century. Somewhat before that, at the turn of the twentieth century, newspaper obituaries were increasingly becoming a critical tool in propagating news of a recent death beyond the immediate neighborhood and to other cities and towns, facilitating communications about the passing of family and friends.[46] Before telephones became common, telegrams also were a customary means of informing family and friends who were further afield. Regardless of how news of a death was received, formally giving condolences in person to the family of the deceased was (and still is) an important obligation for all family, friends, and neighbors. Not doing so, no matter the reason, could be seen as a serious affront. This made the announcement of the death, sonically or otherwise, symbolically very important.[47]

Embodied Grieving and Professional Mourners

> Al-Mu'adida ti'adid wa kul hazinah tibki bukaha.
> The professional mourner laments, and all who are sad cry for their own reasons.
>
> An Egyptian proverb[48]

Public mourning for Egyptian women, more than for the men, was an entirely embodied process. For grief was not shown merely by shrieks, cries, and lamentations, but in varying ways the entire body was involved. In the Egyptian countryside and in popular urban areas, the swaying of the body during the chanting and crying can be likened to a melancholic dance, and traditionally, women used to loosen their hair, dirty their clothes with sand or dirt, and sometimes even rip their garments and hair as an ultimate demonstration of grief.[49] During funerary processions, some mourning women painted their faces, arms, and clothes with indigo dye as a visible sign of mourning.[50] Striking one's own face in loud and audible slaps is common on these occasions. Such repeated slapping has a specific name (*latm*), still used in the Egyptian vernacular, signifying the continuing relevance of this act. Typically, these demonstrations of grief continued for at least forty days, the traditional period of mourning. Public mourning also intensified, of course, during the elaborate, slow-moving funerary processions taking the body from the home to the church or the mosque and then to its final resting place in the cemetery. In the past, music was also an integral part of the funerary soundscape, as Hasaballah bands or traditional musicians played somber tunes in many of these processions. A contemporary observer noted in 1912 that Cairene funerary processions were often "preceded by music, sometimes a band, sometimes only a drum or tom-tom with singers."[51]

Coptic Christian and Muslim funerary practices, laments, and mourning rituals were very similar; with the Coptic practices distinguishable only by the presence of the cross and readings from the Bible instead of the Quran.[52] For example, as Edith Louisa Butcher (1854–1933), a longtime resident in Egypt, and a historian of the Coptic Church astutely observed in 1910:

> Both Copts and Moslems are alike in the frenzied demonstrations of grief which they encourage and indulge in on the occasion of a death. The women dye their hands and faces with indigo, they rend [tear] their garments, and let their hair stream loose and disheveled. Hired mourners add to the clamour by the beat of their tom-toms and the long shrill cries of wailing for the dead.... Before the bier come the male relatives and friends of the deceased; after the bier come the wailing women and all the female mourners. Their cries mingle with the chants and hymns of the men in front. The procession, if Moslem, halts at the mosque for the service...; and if Christian, the body is sometimes taken to a church for the first part of the funeral service.... At the grave, instead of the address to the dead chanted by the Moslems, the Christians read passages from the Gospel.[53]

In Egypt, professional female mourners were (and, to a much lesser extent, are still) hired in order to sonically enhance and amplify the sounds of mourning. There are, however, different types of professional mourners, with varying roles in the funerary process. The professional mourners described by Butcher are called *nawadib* (sg. *nadiba*), and in Egypt they are distinct from another group of female mourners, who are called *mu'adidat* (sg. *mu'adida*). As illustrated in Butcher's description, the performances of the *nawadib* were highly physical and embodied and their primary purpose was to loudly and publicly project the cries and wails for all to hear. Unlike the lamentation known as the *'adid*, which is more subtle and subdued and often composed of longer poetic verses, the mourning expressed through the *nadab* consists of short nonverbal and verbal wails, screams, and declarations. These wails are loudly sounded out, immediately after death, during the funeral procession, or when the body is interned in the tomb (see Figure 5.4). Both the *nadab* and the *'adid* are also often "performed" by family members and neighbors and not just the hired *mu'adidat* and *nawadib*.

The verbal *nadabs* are hardly complex, as they consist of two- to three-word declarations of grief repeated throughout the procession or at the burial site. Examples of *nadabs* include declarations of the general misfortune of the death, with shouts of "my calamity" (*ya lahwi*), "my catastrophe" (*ya musibti*), "my helplessness" (*ya hosti*), and "my ruin" (*ya kharabi*). Husbands or fathers can be loudly described as "my lion" (*ya asadi*) or "my camel" (*ya gammali*), alluding to their economic role as breadwinners. Declarations of love for lost sons and daughters may include shouts of "my darling" (*ya danaya*), "my soul" (*ya ruhi*), "my love" (*ya habibty*), "he was young" (*kan sughayar*), or "I want to die with her" (*Aruh ana wayaha*)![54] These shouted declarations served to focus the energy of the entire funeral toward mourning the deceased. In recounting the many funerals he witnessed in Cairo's Sayyida Zaynab district in the 1920s and 1930s, Fathi Radwan describes the *nawadib* as "actresses or artists who chant away their tears." Unlike many of his contemporaries, however, he does see a purpose to these histrionics, as he understands and appreciates their role in facilitating the mourning process by serving as loud grief-leaders and "inducing tears from even the driest of eyes."[55]

The 'Adid Lamentations

> O daughter cry over your father . . . No one else was as good to you, even your husband or brother
>
> . . .
>
> No one is as dear to a woman as her father . . . He asks about her, and guarantees her fair share
> No one is as dear to a woman as her kind father . . . He asks about her whereabouts

> O kind father, he protected me within his breath . . . He knew how I felt before words were spoken.
> O kind father, he protected me within his shadow . . . He knew how I felt before any told him.
>
> A lamentation by a woman for her father[56]

The poetic ʿadid lamentations are more subdued, somber, and musical than the brief shouted nadabs. Both the nadab and the ʿadid can take place in the same grieving session. As ʿAbd al-Halim Hifni, an Egyptian folklore scholar, has personally observed: "In exceptionally sad occasions as after someone is killed or a young person dies, the ʿadid alone is not enough to express the overwhelming grief felt by the mourners, so they often switch tracks to nadab for a while and then return to the ʿadid."[57] The function of the nadab is to permit a sort of angry cathartic scream that helps relieve the immediate pain; whereas the ʿadid is a more deliberate therapeutic remembering and eulogizing by enumerating the merits of the person who passed away. Toward this end, the ʿadid is composed of longer, rhyming lines with various specific themes, sung in long, regularly held grieving sessions, at times lasting for the full forty-day mourning period and in some cases even longer.

According to Elizabeth Wickett, in the 1980s the typical duration of an Upper Egyptian funeral was seven days, where "lamentations would take place every

FIGURE: 5.4. Funeral Procession Nearing the Cairo Cemetery (ca. 1910).
SOURCE: Douglas Sladen, Oriental Cairo, 1911.

day," and after that they would also take place "on the fifteenth and the fortieth days as well as the year anniversary of the death." All the women of the village were required to attend and participate twice daily at these weeping and lamenting sessions.[58] As Wickett elaborates, "when there is a *wajib* (literally 'duty,' amongst women in Luxor, the word for funeral is seldom uttered) it is incumbent on every woman to come to the funeral on hearing the cry. When each arrives, the gathered throng announces her arrival with a lament and she becomes integrated immediately into the collective mourning."[59] In fact, before the physical arrival of the new mourners, the sounds of their *nadab* screams reach the gathering, who respond back with their own *nadab*, and only then do they return to performing the *'adid* along with the newcomers.[60] These mournful grieving sessions are very much dependent on the group participation and on the multiple voicing and singing by all the women present at the gathering. This involves an elaborate and improvised dialogical interaction among the leading and experienced *'adid* participants, with the rest of the group repeating the key choruses of the chosen laments.[61] Usually, it is the female family members and close members of the community who participate in these funerary songs and dialogues; though often, professional *mu'adidat* are hired to help lead the chorus of laments as de facto grief leaders.

There are a large number of basic lament lyrics, fitting the specific circumstances of almost any funeral. Like the traditional wedding songs, these laments were not written, giving the performers the expected improvisational freedom to adapt the chants to the local situation. In her study of Upper Egyptian *'adid* lamentations, Wickett identified forty-three contemporary lament themes in just Luxor and its surrounding villages.[62] The types of laments chanted for the passing of a young child or a teenager were, of course, distinct from others that were sung for an older matriarch or patriarch. Fathers, mothers, brothers, and sisters had their own types of *'adid*. There were also specific laments for drowning victims, for those who died childless, and even for scorpion bite victims.[63] The lyrics of these laments were never exact and there was a great deal of variation from town to town and even from family to family, though some of the major themes remained the same. There were also broad regional differences, not just for the lamentations but also for other death rituals. A well-known Upper Egyptian lament specifically references this regional difference. "The Stranger's Lament" ("'Adid al-Gharib"), with many improvised variant forms, was sung whenever an Upper Egyptian passed away as a stranger, far from his or her village. The repeating tension in this lament is the certainty, in the minds of those mourning the deceased, that their loved one's funerary arrangements were not properly taken care of, as they would have been back home.[64] Many versions of "The Stranger's Lament" refer to an Upper Egyptian dying somewhere in Lower Egypt, which was considered almost a foreign land in terms of funerary rites:

> The people at the grave [in the delta] do not proclaim the mourning
> Nor do they pull down the lock of hair to the eyebrows....
>
> They do not proclaim her worth, the people at the graves,
> They do not proclaim her worth
> Nor do they wrap up the elegant one in twice-folded cloth.
>
> They do not proclaim her virtues, the people at the graves,
> They do not proclaim her virtues.[65]

Most of these songs, chants, and lamentations helped to relieve the pain of the living for the loss of their loved ones, but they also syncretically preserved an ongoing pre-Islamic and pre-Christian belief that mentioning the names of the deceased and remembering them, facilitates the passage of those who have passed into the afterlife.[66] As Wickett has shown in detail, "women's cultural and ritual practices to expiate grief lie outside the bounds of Islamic or Christian propriety," and that is part of the reason why both Muslim and Coptic Christian clerics continually attack most of Egypt's traditional funerary rituals.[67]

Undertakers and Professional Male Mourners

Though less well known than their female counterparts, during the first half of the twentieth century, there were also professional male mourners for hire, who would lead the entire funerary procession.[68] Although these professional male grievers sometimes somberly chanted religiously inspired condolences such as "God is alive," "we come from God and unto him we return," and "there is no might nor power except in God," they remained silent during most of these processions.[69] It appears that their primary role was more visual than auditory, although they also contributed an olfactory component to funerary marches as they usually carried lit incense burners, swaying them back and forth as they walked. In theory at least, these male mourners were supposed to be impeccably dressed in distinguished attire, lending class and respectability to the entire procession. Sometime in the late nineteenth century, it became popular for some of these men to wear western-style frock coats and shoes. Fathi Radwan, for example, described professional male mourners in 1920s Cairo as wearing "black western style frock coats, carrying with them lit incense emanating the smells of Indian or Javanese herbs and aromatics" (see Figure 5.5).[70] Radwan, however, was much less sympathetic toward these male mourners than he was toward the *muʿadidat* and *nawadib*. He didn't see much purpose in their presence, claiming that most of the men who took on that profession were drunkards or drug addicts, and he incessantly ridiculed their clothing, pointing out their often disheveled appearance:

> In time, however, these distinguished black frock coats lose their color and fall apart just like the men who are wearing them. You can barely see the eyes of these men as they are practically shut due to all the drugs and sleeping pills they abuse. As for their shoes, they represent a pinnacle in high fashion, with the toes often sticking out. Yet, these are the men that are hired to add sophistication and refinement to funerals that are lacking in both. . . . Their incense burners likewise lose their silver sheen and become a hodgepodge of colors as they meekly swing them back and forth. If a foreigner saw this sight, they would assume that these poor people were related to the deceased and the tears had swelled their eyes shut and weakened their bodies. Alternatively, they may think that they are the walking dead, resurrected from their graves.[71]

Radwan was certainly exaggerating for comedic affect, as most of these men were simply poor and making an honest living. Other contemporary accounts reveal that some of these hired male mourners were actually blind men with no other means of support.[72]

According to Radwan, except for the music, the undertaker in the Sayyida Zaynab district "took care of all the funerary services from washing the bodies of the dead to the funerary procession and the burial itself." Undertakers (*hanutiyya*, sg. *hanuti*) also contracted the professional male and female mourners and the Quran reciters for the procession and the funeral.[73] As Radwan recounts, the same music vendors who provided wedding and celebratory music provided funerary music as well: "Right next to the funeral home there is an office specializing in local music bands and performers that are hired for weddings and funerals alike. These bands play happy celebratory songs outside the door of the groom during the wedding night and they perform somber music in funeral

FIGURE 5.5. Two Professional Male Mourners (ca. 1900).
SOURCE: Glass magic lantern slide in the author's collection.

processions, especially for those who died young."[74] Here, Radwan was alluding to the fact that if the person died young and/or the death was an unexpected accident, the level of grief and its outward auditory manifestation through music, and the intensity of the cries and shrieks, might be affected. In fact, in some cases celebratory *zagharit* may be heard if an especially old or pious man or woman has peacefully passed away, as this person is considered by many as someone who is blessed for "living and dying well."[75]

The Economic Acoustics of Death

> *Hagtayn ma hadd yi'lam bihum: mawt al-faqir wa ta'ris al-ghani.*
> Two things no one knows about: The death of a pauper and the debauchery [pimping] of the wealthy.
>
> Egyptian proverb[76]

As this proverb bluntly makes clear, the size and intensity of the expression of grief at a funeral was often proportional to the wealth of the person who had passed away. As with weddings, there is an obvious class dimension to the size of the events and the expenses devoted to funerals, as families of means displayed their wealth during these occasions.⁷⁷ The funerary processions of the wealthy were much more elaborate and costly, as generously feeding the poor was an integral part of the ceremonies:

> If the family is wealthy and important, the funeral procession is as follows: First of all come the live oxen, or other animals which it is intended to sacrifice at the grave for the benefit of the departed soul, or, as the Christians would say, to be given to the poor. Then come camels, loaded with boxes full of bread for distribution. Next come the *fokaha* [*fuqaha'*], or, in the Christian procession, the priest, preceded by the sexton carrying a large silver cross.⁷⁸

The 1911 Cairo funerary procession recounted here was of course atypical, as few could afford to sacrifice oxen, but sacrificing a sheep to feed the poor was more common. The wealth of the family also played a part in how loud and lengthy the mourning period turned out to be. This was due not just to the ability to hire a large number of professional mourners for days or even weeks but also to the fact that larger funerals also, attracted larger crowds. Those who could not pay their respects during the funerary procession attended the official funeral services held at the deceased's home. Just as with weddings, those who could afford it erected a *shadir* in the street or alleyway near the family house to accommodate the mourners.⁷⁹

Families of means were expected to have lavish funerals, and until mid-century, it was not unusual for a wealthy family to arrange a funeral with continuous daily grieving, complete with professional mourners and Quran reciters, for the entire forty-day period after a loved one's death. This could be very expensive, as aside from paying for the lamenters and the Quran reciters, during the grieving and lamentation sessions, the family of the deceased was expected to generously provide food and drink to all those who attended.⁸⁰ Hiring specific Quran reciters also had a clear class connotation, as well-known reciters with celebrated voices charged significantly more money, projecting not only class status but also piety for all to hear and see.⁸¹ This was enhanced even further with the increasing use of loudspeakers at funerary events, allowing the voice of the Quran reciter to reverberate throughout the neighborhood. After the customary forty-day mourning period, many families, regardless of wealth or class, continued grieving in an official capacity on every death anniversary, which usually included a visit to the cemetery. Also, during Eid and other holidays, families often visited the cemetery and the sounds of grief and commemoration for all deceased family members continued on. However, as with the ʿadid and the nadab, grave visitations drew the ire of both religious and secular conservatives.[82]

Conclusion: Silencing Weddings and Funerals?

> Our writers and intellectuals have always been critical of the wedding and funerary traditions here in the East, which have become like worms [sus] eating away at the bones of the nation and bringing upon it catastrophe and disaster beyond imagining or description.
>
> *al-Muftah* (February 15, 1906)[83]

General and deep-seated critiques of many aspects of traditional Egyptian society were plastered all over the Egyptian press throughout the first half of the twentieth century. The quotation that begins this conclusion is from a two-part editorial in the monthly literary magazine *al-Muftah*, titled "Islah al-ʿAdat" (Reforming traditions). The long, eighteen-page essay focused its criticism on lambasting traditional birth, wedding, and death rituals in Egypt, while comparing them to the more "civilized" ways of the west. The writer explained that in order to "help cure Egyptian society from these vulgar social plagues," he would describe how various "advanced and sophisticated" European societies live their lives "from the cradle to the grave." The editorial concluded that Egyptians spend a large and disproportional part of their income, often going into debt, on backward, loud, and vulgar traditional celebrations. The writer advises his readers to learn instead from the frugal, classy, and civilized west.[84] With few exceptions, Egyptian fiction and nonfiction also reflected many critical attacks on various elements of Egyptian society deemed incompatible with existing perceptions of modernity.[85]

Although the *Muftah* editorial was written from a secular, pro-European outlook, critiques of loud Egyptian memorials and celebrations were multifaceted and came from a variety of perspectives. During the first half of the twentieth century, such critiques against varying elements of what was considered traditional society came from both the growing forces of a developing and modernizing "orthodox" Islam—which encompassed a wide variety of Islamic reformists (including, later on, the Muslim Brotherhood)—and the modernist, secular-leaning middle classes and elites. Two well-known turn-of-the-century examples of this genre of critical and classist tracts were Muhammad al-Muwaylihi's *Hadith 'Isa ibn Hisham* (*What 'Isa ibn Hisham Told Us*) and Muhammad 'Umar's *Hadir al-Misriyyin 'Aw Sirr Ta'akhurihum* (The present state of Egyptians, or, the state of their retrogression). Although unlike the *Muftah* editorial, neither of these books advocated for Egyptians to adopt wholesale western-style traditions, they were in agreement over what they perceived as excessive spending on both weddings and funerals, and they detested embodiment and public mixing of genders.[86]

'Umar, as we examined in Chapter 2, was especially cruel in his depiction of the Egyptian poor and working class. His description of a "typical" wedding of a poor family is especially condescending. After complimenting the overall sense of shared joy and cooperation between family members and neighbors in the "weddings of the poor," he qualifies that "their weddings" are disorganized and the happiness is derived from their simplicity and unsophistication. 'Umar's infantilizing prose is especially apparent when describing female participation at the weddings: "If the female invitees want to dance or sing, they do so with the mothers of the bride and groom taking the lead. The only thing ruining these cheerfully serene moments is the loudness of their voices and their ignorance and utter inability to moderate their behavior, leading often to excesses and overspending."[87] 'Umar's judgmental description of a typical "low class" Egyptian wedding reveals a great deal about how, in his mind, loudness was associated with ignorance and vulgarity:

> The wedding festivities typically last a few days before the *hinnah* night and the actual 'wedding.' Traditionally, for a few nights before, which they call the gathering [*al-dumam*], there is singing, drums, *mizmar* and a variety of other music until the night of the *hinnah*, when the bride's family celebrate further, based on what they can afford. All this leads to the wedding day *zaffah* when the groom takes his bride from her house to his. It is during this *zaffah* that they appear as their true selves and reveal their low level of upbringing and elevation, which in truth is the embodiment of ignorance and stupidity. A perfect example of this is the [usual] hiring for the *zaffahs* of naked wrestlers and the infamous Awlad Rabiyyah performers who are known for their lowest [most vulgar] form of loose morals [*khala'a*]. Later on in these processions, there are drummers marching on the ground and others on camelback who noisily ring our eardrums with their beating drums.[88]

Concerning both weddings and funerals, the religious establishment certainly agreed with many of these classist arguments, especially regarding what constituted "vulgarity," and what everyone agreed was wasteful spending, though for the clerics, theological concerns obviously predominated. For example, in a 1911 book, Muhammad Bakhit, who would later become the Egyptian Mufti, declared that mourners in a funeral must not raise their voices aloud and should remain somber and silent. He elaborates that "raising one's voice contradicts the purpose and wisdom of walking in a funerary procession, which is supposed to lead to reflections on death and the hereafter."[89] Bakhit then specifically criticized what he saw in Egyptian funerary practices at the time as explicitly un-Islamic. He specifically rebukes the common Egyptian practices of mourning, including the *'adid*, the *nadab*, and singing aloud during funerals.[90] Even Quran recitations during weddings and funerals, according to Bakhit, should only be allowed under certain sensory conditions:

> Today, people frequently meet to listen to Qur'anic recitation either in their homes, or in the mosques and at times in weddings, funerals and the like. All this is permissible if these meeting occasions are free from what God has forbidden and there is no sonic interference [*tashwish*] with the reciter. The place should also be clean, free from improper licentious behavior, smoking, and foul smells.[91]

These persistent multisensory attacks on popular weddings and funerals continued on multiple fronts. The western-oriented secularists objected to what they perceived as the backward and irrational elements of traditional society, while the Islamic reformists and the growing Sunni orthodox elements in al-Azhar (like Bakhit) were aghast by what they perceived as the non-Islamic or pagan elements that were corrupting "real" Islam. To be sure, these widely divergent groups, despite their obvious ideological differences, agreed on many social and national issues, albeit using different justifications. One of the points of general agreement among both the secular nationalists and those who were more religiously inclined was a disdain for what they perceived as the backwardness, loudness, and "poor taste" of the syncretic practices of the vast majority of Egyptians. It is important to note that the Coptic Christian orthodoxy also shared the general disdain for syncretic, populist mass commemorations and celebrations.[92]

Both the forces of secular modernity and the growth of an equally modern religious orthodoxy, be it Sunni Muslim or Coptic Christian, have been steadily chipping away and eroding some of the colorful and sonorous traditions discussed here. Simultaneously intertwined with these ideological attacks, was the ever-present influence of class and class formation. The growing, educated middle classes, whether they were espousing more orthodox religious or more secular views—or as is more likely, a combination of the two—wanted to distinguish themselves from what they viewed as the backward culture of the masses. Wealth displays, demonstrating generosity, and the ever-present marking of class distinction will always play an important role in most social relations.

Although the overindulgent aspects of Egyptian weddings and funerals were and still are heavily critiqued by both secular and religious voices as being wasteful and even financially ruinous, they continue, albeit aesthetically and musically taking on different and more contemporary forms. Funerals have been arguably more affected by these persistent attacks than weddings, as the *nadab* and *'adid* traditions and professional mourners have been very much in steep decline and are likely to disappear altogether in the near future.[93] The loudness of weddings and the important role of music in them are still features found among all economic classes in Egypt. Usually, what acts as a class marker now is the type of music performed and the choice of venue for the wedding. Adding yet another layer to the evolution of wedding celebrations in modern Egypt are *halal*, or "Islamic," weddings, which started becoming popular at the turn of the twenty-first century. These weddings, which have varying degrees of more conservative music, dancing, and entertainment, might have peaked in their popularity at the time of the 2011 Revolution.[94] To be sure, economic status signaling and class distinction most certainly occur in these halal weddings, but the projection of piety by both families is perhaps just as important, especially as a form of cultural capital.

6
Sounding Out State Power
Cannons, Music, and Loudspeakers

> Wherever Noise is granted immunity from human intervention,
> there will be found a seat of power.
>
> R. Murray Schafer, *The Soundscape*[1]

A CRITICAL COMPONENT of the Egyptian state's efforts at controlling the streets was an ongoing attempt to silence lower-class sounds and "noise." The previous chapters have demonstrated the ways in which street noise was viewed in a negative light, especially when it was associated with the poor and the working classes, as that was perceived as potentially posing a threat to public order. State repression was typically accompanied by varied attacks in the Egyptian media especially targeting popular public expressions ranging from traditional and loud celebrations of weddings, funerals, and circumcisions to religious celebrations of saint's day festivals (*mawalid*; sg. *mawlid,* or *mulid* in the vernacular) and traditional exorcism rituals (*zars*). Most of the condemnations of these popular working-class traditions targeted the loudness, physicality, musicality, and gender-mixing embodiment of these events.[2] Yet, broadcasting street sounds and noise was also an important tool for the Egyptian state, and was regularly used to project power and legitimacy. As this chapter's epigraph suggests, auditory and visual projections of ruling authority have been used in Egypt and elsewhere for thousands of years, and in the modern period, newer tools and technologies are maximizing the state's reach. The modern Egyptian state, especially the khedival and, later on, royal court, has used cannons, gun salutes, fireworks, music, and by the mid-1930s, even strategically placed loudspeakers, in order to sonically project its legitimacy and power into public space. Continuing longstanding practices employed by the Ottomans and Mamluks before them, many of these visual and sonic projections involved the state's appropriation of religious rituals through overt government sponsorship. The biggest of these efforts were the government-sponsored, bi-annual Mahmal pilgrimage parades, the official Mawlid al-Nabi (Prophet Muhammad Birthday) festivities, and the state-sponsored breaking-of-the-fast Iftar tables and nightly celebrations during Ramadan. On a more secular front, loud parades, public celebrations, and/or commemorations associated with key khedival and royal family visits and royal anniversaries, births, weddings, funerals, and coronations were ostentatiously visible and audible in the Egyptian streets.

Booming sounds and noise were created by the Egyptian state's use of cannons, fireworks, music, and loudspeakers in official street celebrations and commemorations meant to make an impression on the amassing crowds, whose sheer numbers must have added a sonic-baseline to the reverberating din. The state's deployment of these sonic tools to broadcast its power and legitimacy was an integral part of the overall public spectacle, intended to impress ears as well as eyes.[3] These sounds were deployed in both the secular and the religious celebrations and commemorations sponsored by the state. In fact, as I discuss near the end of the chapter, despite the persistent marginalization and even repression of many popular religious rituals and festivals, the Egyptian state actively sponsored and appropriated some of these same celebrations in its ongoing attempts to acquire both secular and religious legitimacy.

The State's Cannons

> During this past March 22, 1895, the funeral of Ismail Pasha the previous Khedive was memorialized in Egypt. The crowds amassed in great numbers from Cairo Station [Mahatit Misr] all the way to Muhammad Ali Street. . . . Egyptian and British troops lined up all along the streets parading along the long funeral route. The scene was characterized by an orderly and noble air as at exactly 9:00 o'clock the cannons fired in salute of the passing shroud. The funeral procession started from Cairo Station as cannons were firing every minute. The procession frequently stopped along the way, as it headed first to Opera Square. It was led by the police cavalry and twenty camels, each laden with two boxes of food (for the people), and behind the camels six large water buffalos [for slaughter to feed the poor].
>
> *Al-Muqtataf*, April 1, 1895[4]

The firing of cannons and guns to officially honor or salute key government officials or foreign dignitaries was a common occurrence in nineteenth- and twentieth-century Egypt. In Cairo, cannon fire from the Cairo Citadel was regularly used to mark certain celebrations and festivities, most important of course, was its daily use during the month of Ramadan to mark the breaking of the fast. It also signaled the departure of the Mahmal pilgrimage caravan from Cairo.[5] As the epigraph for this section indicates, the public honoring of the Egyptian ruling dynasty was just as important in death as it was in life. Publicly and loudly memorializing and celebrating births, weddings, and funerals of the khedival (and later on, the royal) family was an essential component of displaying and sounding out khedival power.[6] The elaborate 1895 parade for Khedive Ismail's funeral was especially significant because Ismail had died in Istanbul after sixteen years in exile. He had been deposed by the Ottoman Sultan in 1879, in favor of his son Khedive Mohammad Tawfiq (r. 1879–1892), and spent most of his exile years in Naples and then Istanbul, where he died on March 2, 1895. His body arrived in Cairo three weeks later for this elaborate state funeral. The significant time, money, and resources devoted to bringing his body back to Egypt and arranging this expensive funerary parade and ritual was deemed necessary by his grandson, Khedive Abbas Hilmi II (r. 1892–1914). This was not done merely for sentimental reasons but was also clearly political, to demonstrate the power of the khedivate, and to clearly and symbolically reaffirm its roots in Egyptian soil.

The 101-gun salute during Ismail's funerary procession, and the timed precision of the cannon fire, starting at exactly 9:00 a.m. and discharging every minute, added a further sonic dimension to the visual and auditory orchestration of the marching infantry and cavalry (see Figure 6.1). As was the case for most of the state parades, part of the power they projected came from the simultaneous sight of orderly and uniformed troops marching in unison, coupled with the sounds of the hooves of the trotting horses and the stomping of hundreds of troops marching in step. It was indeed a visual and auditory feast, meant to impress the power of the state on ordinary people. For Cairo's poor, it was also a literal feast, as meat from the six large water buffalos in full view of all would soon be distributed to those in need.[7] Large, open city spaces, like Opera Square (also called Ibrahim Pasha Square), where Ismail's funerary procession stopped for a while, were ideal locations for experiencing these auditory and visual spectacles. These squares became de facto massive stages for the unfolding live drama and

FIGURE 6.1. The Funerary Parade of Khedive Ismail at Opera Square (March 22, 1895).
SOURCE: Fredrick Penfield, Present-Day Egypt (New York: Century, 1907).

allowed large numbers of spectators to see and hear the processions. The obvious and imposing visual presence of the statue of Ibrahim Pasha (Ismail's father) on horseback, right in the middle of the square, added yet another layer of aesthetic symbolism, reinforcing the dynastic claims of the Muhammad Ali family.

The firing of a 101-gun salute for important khedival ceremonies continued for decades, even after Egypt briefly became a sultanate in 1914 and then a kingdom in 1922, when the country was granted nominal "independence" from Great Britain. For example, with every convocation of a newly elected parliament, starting with the first on March 15, 1924, the king and his entourage made the short carriage drive from Abdin Palace to the parliament building with the same pomp and precision as was displayed in Khedive Ismail's funeral. Thus, during the morning of March 15, as the pro-British *Egyptian Gazette* reported: "At 9:45, the thunder of guns was heard in the house, as in all the city, announcing that King Fuad had left Abdin Palace to open the Parliament." The 101-gun salute was still firing as the king's royal carriage traveled in a long procession, escorted by squadrons of mounted police and royal bodyguards. The visual and sonic royal choreography was precise, as "at exactly ten o'clock," while the guns were still booming, King Fuad entered the halls of parliament "wearing the uniform of the Commander-in-Chief of the Army."[8] According to J. Morten Howell, the

U.S. consul general in Cairo, during these convocation of parliament ceremonies, "the same royal salute of 101 guns was fired in the cities of Alexandria and Port Said at the same hour."[9]

Royal Guns and Festivities in the Nation's Periphery

Though projecting state power in the capital city and in large city squares provided for an impressive multisensory spectacle, this technique was arguably more important in the nation's periphery. In fact, when funds were available, royal visits, births, weddings, and coronation anniversaries were locally celebrated with elaborate festivities in which small-town and village elites and representatives of the king (if not the king himself) were present before the entire town. The firing of gun salutes and the playing of the "Royal Anthem" in honor of the king or his representatives was an essential component of these celebrations, and this sonic accompaniment usually began at the town's rail station and continued as the royal entourage paraded through town on its way to the celebratory venue. A classic example of such an event was detailed in a lengthy descriptive editorial published on October 18, 1925, in *al-Bahlawan* (The acrobat), a monthly satirical periodical published in Shibin al-Kum, the capital of the Minufiyya province in the Nile delta. The editorial provides a detailed description of a local provincial celebration of King Fuad's eighth anniversary as ruler of Egypt.[10]

The editorial, titled "Celebrations of the Coronation of Fuad I as ruler of Egypt in Shibin al-Kum," describes how "on the morning of October 9 [1925] thousands upon thousands gathered in town, including the village chiefs ['*umad*], notables, and regional government employees." The entire town was elaborately decorated for the occasion, with hundreds of Egyptian flags and banners, and "when night arrived all eyes were gloriously impressed by brilliant electric lights." The festive lights and decorations were supplemented by "loud chants supporting the king of Egypt and his heir" and the "ululations [*zagharit*] of the women were loudly heard." The editor, who publicly at least, was an enthusiastic supporter of the king, hyperbolically continued: "it was as if there was a holiday celebration in every house, a wedding in every street, a gathering in every mosque and a celebration in every square."

The celebration's main event took place in a large pavilion tent (*shadir*) set up for the occasion, as the invited guests sat "on the hundreds of gold colored chairs" rented for this occasion. At 10:00 a.m., the king's official regional representative for this event, "Ibrahim Bey Rushdi Qamha, the deputy *mudir* of Minufiyya, arrived in his festive motorcade, and was greeted with a cannon fire salute and the playing of the royal anthem." After the morning speeches and festivities, "food and alms were distributed to the poor and needy." In the evening, the opulent pavilions were "lit up with electric lights, and government employees, village chiefs, workers, and merchants attended" the continuing celebrations. According to *al-Bahlawan*, the festivities, which included music, speeches, and Quran recitations, continued until dawn. The editor made sure to mention at the end of his commentary that as the attendees were leaving the celebration at dawn, they "chanted in salute of his highness the King and his heir Prince Faruq." Throughout the khedival and monarchal period, events like these were commonly held in the countryside, in some of the smaller provincial towns in the delta, and in Upper Egypt. The entirety of the aesthetic, acoustic, and ritualistic dynamics of these events, from the opulent displays, the lights, music, and cannon fire to the distribution of food to the poor, was meant to sensorially display royal power and legitimacy.

The Sounds of the Ramadan Cannons

Throughout the nineteenth century and the first half of the twentieth century, cannon fire was associated with royal power and gun salutes. However, the booming sound of the Ramadan cannon is the sound more connected to the hearts and minds of ordinary Egyptians, and until this day, it is still directly associated with the month of fasting. By the second half of the nineteenth century, the expectation of most Cairene Muslims was that they would hear the booming Ramadan cannon before breaking their fast. This became a part of the ritual of Ramadan itself, as a turn-of-the-century observer remarked upon seeing the many Cairenes who "stand at the doorway to listen for the discharge" of the cannon.[11]

By providing cannons and the gunners to fire them during the month of Ramadan, and by regulating and synchronizing the daily time of the firing, the Egyptian state sonically signaled its daily commitment to sponsoring the Ramadan ritual during the entire month of fasting.[12] The practice of firing of the Ramadan cannon twice daily—to signal the beginning of the fast in the predawn hours and the breaking of the fast after sunset—during the entire month of Ramadan started in Cairo perhaps as early as the fifteenth century.[13] It is unclear whether this practice continued with regularity for the next three centuries until the advent of the Muhammad Ali Dynasty in 1805. However, by the late nineteenth century, if not earlier, the state practice of firing the Ramadan cannon extended beyond the Citadel guns in Cairo to Alexandria and to other major cities and towns in the delta and beyond.

For example, an 1898 telegraph from the Shari'a Judge of Port Said to the office of the Khedive Abbas Hilmi II inquires about the reason for the silence of the Port Said cannon during the first few days of the month of Ramadan. The message indicates the expectation of hearing the cannon fire and the clear association of the cannon fire with the breaking of the Ramadan fast.[14] The judge begins his telegraphed petition by invoking all the religious legitimacy he can muster, declaring that "respect for religion (& religious tradition) is a moral duty." He reports that "Ramadan is already here, yet the cannons have not fired as they were supposed to . . . in all Islamic cities." As if to legitimize his grievance, he adds that because of the silence of the Ramadan cannon, "Muslims here of all nationalities were in a state of unease." Cleverly though, the Port Said judge offers a possible solution and volunteers that, "if the reason for the [silence of the cannon] is the unavailability of gunners and cannon crews, the (Port Said) Port Police are well capable of operating the cannon." The message ends with a final plea: "On behalf of myself and as a representative of the majority of Muslims who live here, I humbly request your royal highness to give the permission to fire these guns during the appropriate times."[15]

It was not just cities and major towns like Cairo, Alexandria, and Port Said that expected the daily firing of a cannon during the month of Ramadan. In fact, because of the increasing national association of the booming sound of the cannon with breaking the Ramadan fast, smaller towns and communities throughout Egypt petitioned the state to provide them with a cannon and presumably gunners and munitions to operate it throughout the month of Ramadan.

For example, on July 30, 1914, the residents of the villages of Mit Ghamr and Ziftah—located in the Nile delta, just east of Tanta and approximately 80 kilometers north of Cairo—sent a petition to the Egyptian Ministry of War requesting the placement of a cannon between their two villages for the entire duration of the month of Ramadan.[16] As these towns are located across the Damietta branch of the Nile River from each other, the petitioners estimated that one cannon, placed by the river, was more than enough. Given that the centers of the villages are only about 1 kilometer apart, the sound from the cannon would have easily and loudly been heard in both villages and would most likely resonate for several kilometers beyond. This request of course required quite a bit of expense and logistical support from the state. Aside from transporting the cannon to the desired location, the army would need to supply the ammunition and the necessary gunners and artillery crew to fire the cannon twice daily for a month.

The cost would also include the feeding and housing of the gunners and artillery crew. The response from the Ministry of War was swiftly forwarded to the Council of Ministers (Majlis al-Nuzzar). Aware of all the necessary logistics, the letter declared that it was possible to provide a cannon for limited time during Ramadan, "as long as this request does not incur any expenses on the Ministry of War."[17] It is not clear from these reports whether the notables of Mit Ghamr and Ziftah were able to provide the funds to support the placement of the cannon between their villages during Ramadan. However, what is evident from this story and others like it, is that the cultural expectation of having a cannon sound off during Ramadan was becoming nearly universal in turn-of-the-twentieth-century Egypt.[18] Fulfilling this expectation was almost entirely a state operation, as only the military had access to cannons.

Although gun and rifle fire was used by ordinary citizens, especially in the countryside and in Upper Egypt, to sound out wedding celebrations, providing cannon fire was the exclusive purview of the Egyptian authorities and was an integral part of sonically projecting state power. The Ramadan cannon, in contrast, and the thunderous sound it projected, became synonymous with Ramadan as a holiday. In fact, throughout most of the monarchy period, as part of the Eid al-Fitr celebrations, marking the end of the month of Ramadan, the Egyptian army fired a cannon placed in the large parade grounds in Abbasiyya's al-Rasdikhana Square (later called al-Ghafir Square) (see Figure 6.2).[19] Starting as early as the 1920s, photographs and drawings of the Ramadan cannon, along with the crescent moon, and the lit Ramadan lantern, symbolically represented Ramadan in the Egyptian illustrated press in both editorials and advertisements. Furthermore, these images naturally migrated to greeting cards, television, web pages, and social media. The booming sound of the Ramadan cannon is still relevant, and holds some emotional resonance for most Egyptians, though many hear it only in mediated forms, through the speakers of their computers, smartphones, radios, and TV sets.[20]

State Music and the King's Brass Bands

> By company or regiment, soldiers are so frequently marched through the streets that the visitor might believe Cairo to be a vast military camp. Martial music is the adjunct of every function and every anniversary, religious and festive. Drum and fife corps, full military bands, some of them mounted, parade daily, playing

frequently the beautiful Khedival Hymn. It is a part of the scheme of administration to keep the soldier in evidence, impressing the simple native with the importance of the army, in which he must serve, however reluctant.

Frederic Penfield, *Present-Day Egypt* (1903)[21]

As I discussed in the last chapter, military-trained Egyptian brass bands played mostly Western martial music, like the "Marseillaise" and dozens of other pieces of military-style marching music, for the army and often for the wider public during official parades and public processions.[22] The "Khedival Anthem" was of course also a regular part of their musical repertoire in official ceremonies and occasions. By the mid-nineteenth century, in urban areas at least, Egyptian ears were certainly accustomed to regularly hearing military brass bands as they also performed in the streets and squares during official state celebrations and in many of Egypt's public parks, including the Azbakiyya Garden and later on even the Cairo Zoo. Azbakiyya Garden had a dedicated "bandstand" next to its outdoor theater, and Egyptian military bands regularly played there.[23] Another major bandstand was located in Muhammad Ali Square in Alexandria, and it remained there until the 1950s. After the 1882 British occupation, the "bands of the British regiments in garrison" took turns

FIGURE 6.2. Egyptian Army Cannon Used to Announce the End of Ramadan (Eid al-Fitr).
SOURCE: *Al-Ahram*, January 17, 1934.

playing there in the summer, including the bagpipers of the Scottish Borderers playing Highland music.[24]

In the Giza Zoo, which opened in 1891, an Egyptian military band regularly played, providing entertainment to the thousands visiting there every year. The number of visitors would increase even more after the Giza pyramids tramway, which passed right by the zoo's main entrance, was finished near the end of the century.[25] By the late nineteenth century, a variety of military-style Egyptian and British brass bands became a regular part of the urban soundscape.[26] For example, this is how a 1911 visitor to Cairo described the entire Azbakiyya district: "The Esbekiya [Azbakiyya]—I refer to the quarter of sin, and not to the fainéant garden—was braying with brass bands, blazing with flares and electric lights, buzzing with people, having a night out."[27]

Naturally, Egyptian military brass bands did not play only in public entertainment venues. In fact, the state events at which cannon salutes were used also had accompanying music; even during Khedive Ismail's funerary procession, military bands played somber funerary hymns. Most of Egypt's military bands played music in all kinds of official state events, from parades to celebrations, jubilees, and official state visits by ambassadors and foreign leaders, where they played the national or royal anthems of the various nations represented at these events. The pomp and elaborate ceremonial rituals of diplomatic visits had a distinct soundtrack that was heard by all within a few city blocks. Thomas Harrison, who served as the American consul general in Cairo from 1897 to 1899, vividly describes his first official visit to Khedive Abbas Hilmi II in Abdin Palace and, in the process, captures his impressions of many of these sights and sounds. According to Harrison, the diplomatic festivities were set into motion by the arrival of the khedive's grand master of ceremonies to his residence. The master of ceremonies was sent to pick Harrison up in what he described as an elaborately decorated gala coach, "with gold galore on the outside, body and gear all being brightly gilded." The coach was escorted by at least fifty cavalrymen, "who sounded bugles" announcing the slow "au pas" march to the palace.

The police emptied the entire route for the fifteen-minute journey, and spectators lined each side of the road. The sonic climax of the short journey must have left an impression on Harrison as he elaborately describes the moment of his arrival at Abdin Palace: "The band of an Egyptian regiment . . . struck up 'Hail Columbia' . . . and as I descended from the coach and entered the Palace, the guns from the citadel boomed loudly a 'salvo' of twenty-one resounding guns upon the still, dry air, which were echoed back again and again from the near-by Macadam [Muqattam] Hills."[28] This scene was not unique and was frequently repeated throughout the long nineteenth century. Automobiles and motorcycles with their loud sirens would soon replace the coach and cavalry with their bugles, yet the ceremonial purpose of impressing the power of the state on the foreign dignitaries, and more importantly on the Egyptian masses listening and observing on the side of the road, remained.

Egyptian military bands most frequently played the "Khedival Anthem." This piece, which was also called "Salam Afandina" (the Khedive's salute) and which did not have any lyrics, was commissioned under Khedive Ismail in 1869, and despite the fact that it was composed to exclusively honor the khedive, it became the de facto Egyptian national anthem for almost a century, from 1869 to 1958. In 1919, a contemporary observer remarked that "all Egyptian and British bands in Egypt and the Sudan play this melody upon the arrival and departure of the Sultan on state occasions."[29] As the name of the Egyptian state changed, the name of the "Khedival Anthem," also changed, becoming the "Sultani Anthem" (1914–1922), the "Royal Anthem" (al-Salam al-Malaki) (1922–1953), and for the first five years of the Egyptian Republic, the "Republican Anthem" ("al-Salam al-Jumhuri").

The "Royal Anthem" and Egyptian military band music would soon be played repeatedly on Egyptian radio for thousands of listeners to hear, multiplying its reach through the power of mediation. Since its inception in 1934, Egyptian State Radio has ended its broadcasting every night with one of the so-called King's bands playing a rendition of the "Royal Anthem." Through this mundane, though repetitively effective routine broadcasting, government radio expanded the sonic reach of the Egyptian state well beyond any single "live," publicly held performance or event. On special occasions, marking state, religious, or royal commemorations, most of the station's programming would be devoted to these events. For instance, a significant portion of the radio programing on Thursday March 26, 1936, relayed the celebration of King Fuad's birthday, and the cover of that week's issue of the government-owned *al-Radiu al-Misri* magazine featured an elaborate colored photograph of King Fuad, ceremonial sword in hand and dressed in his official military regalia. The main article in that issue, titled "'Id Milad Sahib al-'Arsh" (The king's birthday), offered a lengthy paean praising not only Fuad's reign but also his entire dynasty, starting with Muhammad Ali.[30] It described Fuad as "a great politician, military-man, ruler, organizer, reformer, builder, agriculturist, originator, businessman, and industrialist," and it ended by shamelessly declaring that Fuad "possesses the best qualities of ten great men and his greatness is humbling to all, as everyone benefits from his genius."

As part of the airtime devoted to honoring Fuad, a military band played a variety of celebratory tunes throughout the day, and a photograph of the band was featured in the magazine (see Figure 6.3). The caption read: "One of his Highness the King's bands, may God protect him and grant him long life."[31] The band's first live concert was, according to the magazine, especially "ordered by his highness the King" and played from 9:00 a.m. to 10:40 a.m. Readings of odes and panegyrics to the king were interspersed throughout the day, along with more music honoring Fuad. The state's use of multiple media outlets to legitimize royal sovereignty and validate its authority by strategically using its own music and broadcasting its own sound, extended to the religious sphere, as projecting piety was just as important as projecting power. To what degree these techniques were effective is a difficult question to answer, as listeners could very easily tune out, literally and figuratively.

FIGURE: 6.3. The King's Band.
SOURCE: *Al-Radiu al-Misri*, March 21, 1936.

Sounding Out Piety: State Appropriation of Religious Celebrations

At the turn of the twentieth century, hundreds of carnival-like, independent saint's-day festivals (*mawalid*) took place year-round throughout Egypt. According to Ali Pasha Mubarak, in the late nineteenth century over eighty *mawalid* were celebrated annually in greater Cairo alone.[32] These local Cairene *mawalid* were in addition to the larger festivals that were also celebrated elsewhere in Egypt and the rest of the Muslim world. Most villages and towns of reasonable size had at least one *mawlid* for a local saint that the people celebrated yearly. Though most of these *mawalid* were Muslim, some were Christian or Jewish, and they were by and large celebrated by all, as most served economic, social, and entertainment purposes. An Egyptian *mawlid* is a loud and festive carnival of lights and sounds. Although devotional Sufi *zikr* (chanting) and dancing are integral to a *mawlid* and practiced by some, more secular carnivalesque entertainments and pursuits predominate. It is also a traditional fair, complete with food stalls, music, fireworks, makeshift circuses, carnival rides, Ferris wheels, magicians, farcical street theater, male and female dancers, acrobats, snake charmers, wrestlers, prizefighters, puppet shows, and more.[33] Within this multisensory feast, public gender segregation norms are relaxed, and to some extent, social hierarchal differentiation as well.[34]

Such challenging of social hierarchies and orthodoxies and subverting of authority, even temporarily, was threatening to many. As Samuli Schielke and Meir Hatina have accurately shown, intense criticism of and attacks on *mawalid*, Sufi practices, and other forms of popular Islam occurred throughout the nineteenth and twentieth centuries.[35] By the 1940s, many of the smaller *mawalid* had died out, as state supervision increased and prior permission from the Ministry of the Interior was required annually.[36] Although orthodox religion, the Egyptian state, and secular modernizers ostensibly occupied different bands of the ideological spectrum, they all represented centralizing and growing forces that attempted to homogenize the diversity of beliefs and practices in popular religion, in order to "uplift the ignorant and superstitious masses." This more rigid, intrusive, and institutionalized view of public morality, taste, and aesthetics steadily worked to marginalize and silence some of these popular practices. As I examined in the previous chapter, many aspects of traditional weddings and, especially, funerals were attacked in various ideologically diverse critiques, though these attacks pale in comparison to those directed against independent popular *mawalid*, exorcisms, and other aspects of popular religion.[37] Most of the attacks on these popular traditions featured critiques of the musical, dancing, and general embodied nature of these events.[38] At the turn of the century and beyond, many newspapers, magazines, and journals attacked these traditions as inherently backward and called for their eradication.

A good example of this sort of anti-*mawlid* modernizing discourse was on full display in a cartoon printed in the satirical entertainment magazine *al-Ithnayn wa al-Duniya* (see Figure 6.4). The angry turbaned man pointing expressively with his left hand is Muhammad al-Ghunaymi al-Taftazani, a reforming Sufi shaykh of the early twentieth century, who crusaded against "harmful" innovations (*bidaʿ*) in *mawalid* and other Islamic rituals and celebrations. In the cartoon caption, al-Taftazani yells out: "O people, this is supposed to be the commemoration of the Prophet of God, who provided for you divine guidance. It is shamefully forbidden for you to deface [or dishonor] the glory of his blessed *mawlid* with these shameful embarrassments, which Islam is innocent of."[39] The images surrounding al-Taftazani depict some of the people and acts that he found objectionable. They include a snake charmer, a magician, a dancing monkey and its trainer, faith healers, and a female dancer or singer and her band of musicians. According to Muhammad Bakhit, who would become Mufti of Egypt from 1914 to 1921, "the Prophet's Mulid celebrations can be lawful if they were respectful, austere and follow Islamic law by not including "dancing, passionate love songs dedicated to boys and concubines, or any mention of alcohol and the like." Although "all types of fortuitous entertainments and morally corrupting songs" are strictly forbidden, chants that respectfully "sing the praises of the prophet" were permissible.[40] Officially (and ironically) though, it was not the "singing and dancing" that led to government harassment of many of Egypt's *mawalid*. Instead, bureaucrats used health code violations and cited preventive measures against cholera and other potential disease outbreaks to shut down, delay, or harass many of the smaller saint's day festivals, and the Egyptian archives are full of petitions from local notables demanding the reopening of their local *mawalid*.[41] Discourses of rationality and the pretensions of civilizational uplift can be oppressive, and are almost always silencing, if not outright stifling of popular traditions.[42]

FIGURE 6.4. "At the Mawlid."
SOURCE: *Al-Ithnayn wa al-Dunya*, June 25, 1934.

The Government's Sound and Light Show

> The official feast looked back at the past and used the past to consecrate the present. Unlike the earlier and purer feast, the official feast asserted all that was stable, unchanging, perennial: the existing hierarchy, the existing religious, political, and moral values, norms, and prohibitions. It was the triumph of a truth already established, the predominant truth that was put forward as eternal and indisputable. This is why the tone of the official feast was monolithically serious and why the element of laughter was alien to it. The true nature of human festivity was betrayed and distorted.
>
> Mikhail Bakhtin, *Rabelais and His World*[43]

One way to silence popular noise was to appropriate it and drown it out with official state noise. Several yearly state-sponsored celebrations—the most important of which were the official Mawlid al-Nabi (Prophet Mohammad Birthday) celebration and the Mahmal caravan pilgrimage parade in Cairo—served to sonically and visually project the power and benevolence of the ruling regime.[44] The Mahmal carried the valuable and richly decorated, Egyptian-made *kiswa* (black silk cover for the Kaaba), which was annually sent to Mecca before the pilgrimage season. The old cover was brought back after the pilgrimage for yet another government-sponsored parade. These and other state-sponsored religious celebrations were primarily meant to demonstrate state power and gain religious legitimacy from the masses, and unlike other popular mawalid and rituals, were never sanctioned for producing unwanted noise. As R. Murray Schafer spells out, "wherever Noise is granted immunity from human intervention, there will be found a seat of power."[45]

The "power" of the Egyptian state was at the center of the festivities during both the official Mawlid al-Nabi and the yearly Mahmal parade celebrations. The heart of the festivities for both events was an elaborate military parade, complete with marching infantry, cavalry, and artillery and, by the 1930s, even tanks and armored personnel carriers.[46] In his 1911 book on the harmful non-Islamic innovations found in contemporary popular rituals, Muhammad Bakhit, not surprisingly, declares that for the Mahmal and the Prophet's birthday, "military parades and the like are among the permitted innovations (*min al-bidaʿ al-mubaha*)."[47]

Bakhit, who was likely auditioning for his future position as Egypt's Mufti, was very accepting of the official state-sponsored Mawlid and Mahmal celebrations. He treads carefully in his book, so as not to offend the khedive or other high-ranking government officials. For example, after mercilessly attacking the dancing, singing, and gender mixing that takes place in the Egyptian Mawlid al-Nabi and during the Mahmal celebrations, Bakhit makes sure to qualify that "the public places where the religious scholars and the notables assemble along with the Khedive are always proper and only what is lawfully permitted takes place there."[48]

Both of these celebrations were usually attended by the khedive (or his representative), government ministers, notables, well-known religious scholars—including the Mufti and Shaykh al-Azhar—and other dignitaries and public figures. Both the Mahmal and the Mawlid al-Nabi had dedicated official celebration locations within greater Cairo. The main location for the Mahmal parade was the large square below the Citadel and overlooking Muhammad Ali Street, where a large reviewing stand was placed for the khedive and all the notables.[49] The official spot for the Mawlid al-Nabi celebrations was an even larger square in Abbasiyya, next to the army barracks. In 1882, Khedive Tawfiq (r. 1878–1892) had moved the Mawlid location there from Qasr al-'Aini. Until the 1950s, it was a wide-open plot of land bordering the desert that, for most of the year, served as a practice and drilling ground for the Egyptian army. Although this area was in the outskirts of Cairo, since the late 1890s it had been connected to the city via tramway.

As the time for the Mawlid approached, over a dozen large, extravagant, and richly decorated cloth pavilions were built for the khedive and the various government ministries (see Figure 6.5). Next to them were located the pavilions of notables, high-ranking clerics, including Naqib al-Ashraf (head steward of the descendants of the Prophet). For the tens of thousands of ordinary Egyptians who visited the site, it was no doubt awe inspiring. Douglas Sladen, who visited the Abbasiyya Mawlid in 1910, was impressed by the scale and opulence of the spectacle:

> Presently the long line of blocked tramcars showed us that the plot was thickening, and when we suddenly swept round them, we came upon an extraordinary spectacle—a vast rectangular space, about the size of the Stadium, surrounded by enormous pavilions broidered with the most brilliant specimens of the tentmaker's art. Some of them must have been a hundred feet long and fifty feet high; their fronts were open, the flaps being turned up like the starched flaps of a French

nun's coif, flinging to the sunshine and the breezes the gleam of the red and blue and gold in which the texts from the Koran were emblazoned on them.⁵⁰

Sladen had an insider's view of the opulence of these pavilions, as he was the personal guest of Shaykh al-Bakri, the Egyptian Naqib al-Ashraf. Inside al-Bakri's pavilion were "rich deep carpets and scores of easy chairs," some covered in pink satin. Adding to the richness of the large tent, "magnificent crystal chandeliers hung down from the lofty roof;" and "festoons of red and white electric lights were looped all round it."⁵¹

The khedival/royal pavilion was the most impressive of all. The Abdin Palace files at the Egyptian archives are full of details and expense reports for the sponsorship of this annual Mawlid celebration.⁵² Every year, orders from the palace were given out to the Mayor of Cairo who assigned a city engineer to set up the pavilion. A long list of furniture, decorative items, and royal accessories was

FIGURE 6.5. Mulid al-Nabi in Cairo's Abbasiyya Grounds: The Awqaf Ministry Pavilion (1914).
SOURCE: Rare Books and Special Collections Library, American University in Cairo: http://digitalcollections.aucegypt.edu/cdm/singleitem/collection/p15795coll15/id/1273/rec/16.

tabulated and acquired from the warehouses to decorate the pavilion. The items included a large throne-like golden chair, twenty-seven golden chairs for guests, silk, red velvet, Persian carpets, a long tiger skin carpet, and a large ornamental crown set up right in front of the pavilion. Electric lights lit up the large crown like a Christmas tree at night, signaling to all the presence and sponsorship of royal authority.[53]

Unlike the Mahmal festivities, which mainly took place during the day, the Mawlid celebrations went on well into the night. Sladen remarks that at night the Abbasiyya parade grounds were even more impressive as "darkness fell, and the electric lamps flashed out like stars."[54] In another passage, he comments that Abbasiyya was a "splendid spectacle late that night, for the richly decorated tabernacles which surrounded the vast square were a blaze of light, and full of holy men reciting the Koran, and dancing and singing of a religious nature."[55] The bright nighttime electric lights, along with the constant music and "the tum-tuming of drums from various points" as described by Sladen, created what must have been an impressive sonorous spectacle.[56]

As many examples throughout this chapter show, these large state celebrations had a strong and perhaps sometimes overwhelming sonic element, featuring military music, traditional bands, Sufi dancers and musicians, and government-sponsored fireworks, cannon fire, and gun salutes.[57] During the Mahmal, for instance, the first signal that the procession had arrived was the sound of the band playing the "Khedival Anthem," followed by "the guns thundering out," and the army standing and saluting, signaling the arrival of the prime minister or the khedive. Directly afterward, "a burst of Oriental kettle-drums," and Egyptian double-reed flutes (*arghul*) were heard formally announcing the arrival of the Mahmal procession.[58] Egyptian military brass bands were front and center during the official segments of both celebrations, and visually and sonically, they added to the authoritative air of the festivities (see Figure 6.6).[59]

In its drive to obtain religious legitimacy, the Egyptian state appropriated, sponsored, and "made official" some of the major Islamic religious festivities. Yet despite the fact that the state and the Egyptian royal family appropriated elements of these popular religious festivities with their displays of power, light, and noise, the carnivalesque atmosphere continued out of sight, though perhaps not out of hearing range, of the large official pavilions. As alluded earlier, aside from the official government marching brass bands, during both of these religious parades there were dozens of other amateur and professional bands and musicians playing

FIGURE 6.6. Military Band at the Prophet's Birthday Parade in Abbasiyya (January 1949).
SOURCE: ACME News Service, copy in the author's collection.

music and beating their drums in celebration. Many of these bands marched in the celebratory procession along with various Sufi orders. Contemporaries described the sounds of tambourines, bagpipes, "droning drums," and "tinkling cymbals," and incessant singing and chanting continuing through the night, well after the official state celebrations were over.[60]

The State's Loudspeakers and the King's Speech

By the mid-1930s, on nationally or religiously significant occasions, loudspeakers placed in many of Egypt's city squares added to the Egyptian state's sonic arsenal. Loudspeakers changed the dynamics of conveying messages and state propaganda to the Egyptian masses. They were not only used for live events but often also used as an extension of state radio to sound out nationally important speeches, music, and during religious celebrations, even Quran recitations. Indeed, many of the state-sponsored events discussed previously made use of loudspeakers. More importantly for our purposes, loudspeakers dramatically transformed large events, allowing media content—whether broadcast from a stage nearby or from a radio station hundreds of miles away—to loudly reverberate, simultaneously reaching all the people in a large gathering. Allowing the state to reach out and sonically connect to thousands of listeners in large gatherings is an altogether qualitatively different experience, as large-group dynamics come into play, potentially enhancing the emotional connections between individuals in a mass of people as the broadcast speeches and music loudly reverberate around them. It is hardly surprising that in the 1930s, Europe's fascist regimes relied heavily on loudspeakers to propagate their message to the masses.[61]

Egyptian State Radio and publicly placed loudspeakers played an important role in the critical period of transition after the death of King Fuad and through the ascension of the young King Faruq. As soon as King Fuad passed away on April 29, 1936, plans were set into motion to hurry Prince Faruq back from England, where he was preparing to study at the Royal Military Academy in Woolwich. The sixteen-year-old Faruq traveled via ship, escorted by the Royal Navy until reaching Egyptian waters, where his ship was met by "the Egyptian coastguard vessel *Prince Faruq*, and by a squadron of Egyptian army planes." The sea and air escort continued until his arrival at Alexandria Harbor, where "a salute fired by the guns of Fort Saleh and the British Navy" sounded out. A makeshift procession honoring his arrival accompanied him as he made his way to Ras al-Tin Palace in Alexandria and then immediately to the train station, where at 9:30 a.m. he boarded the Royal Train to Cairo. Along the way, short strategic stops were made in the towns of Damanhur, Tanta, Kafr al-Zayat, and Banha, where, according to British reports, "each station was packed with local notables and cheering crowds who gave their young King a rousing welcome." On May 6, 1936, Faruq arrived at Cairo Station at 12:30 p.m., and just two days later, he made his first national address to the nation.[62]

This short speech was delivered on Friday, May 8, and then broadcast on Egyptian State Radio. The king recounted the reason for his brief seven-month stay in England, mourned the passing of his father, and projected his seriousness of purpose in assuming his ruling responsibilities.[63] Being that it was his first radio speech, it was not broadcast live, but was recorded and then reviewed by the king and his handlers before its official broadcast minutes later. The king did not have to leave his palace to record or broadcast live on Egyptian radio, reaching tens of thousands on a moment's notice. The recording took place at the king's desk in Abdin Palace, which was equipped with the necessary equipment. *Al-Radiu al-Misri* magazine placed a photograph of the king, posed during the recording session, on its front cover for its May 16 issue. According to the article, the microphones were placed in two tall desktop wooden boxes, which can be seen flanking the King in the printed photograph. (see Figure 6.7).[64]

By 1936, a number of the king's palaces had been equipped with microphones and radio broadcasting equipment, and for the rest of his reign, Faruq regularly made broadcast speeches from the Abdin and Qubba Palaces in Cairo and, during the summer months, from the Montazah and Ras al-Tin Palaces in Alexandria. When these speeches were made on public holy days or important government occasions, the broadcasts were also played over loudspeakers throughout Egypt.[65]

One of the earliest uses of publicly placed loudspeakers to broadcast an official Egyptian government event took place just two weeks after Faruq's first speech. On the morning of Saturday, May 23, 1936, the Egyptian Parliament's opening ceremony was broadcast live on Egyptian State Radio. Days before the event, the Ministry of Communication, through its department of Telephones and Telegraphs, placed large loudspeakers in several locations in Cairo and other major cities. The entire parliamentary ceremony was then broadcast live on Egyptian radio and simultaneously pumped through these multiple speaker systems set up in various places throughout Egypt, including

FIGURE 6.7. The King Recording His First Radio Speech (May 8, 1936).
SOURCE: *al-Radiu al-Misri*, May 16, 1936.

most prominently in Ibrahim Pasha Square in Cairo (see Figure 6.8).⁶⁶ The highlight of the opening ceremony was the reading of the King's Speech by the newly elected Wafdist Prime Minister Mustafa al-Nahhas Pasha (see Figure 6.9). That day, Nahhas' voice reverberated throughout the streets of Cairo and beyond.

Press reports and letter and telegraph correspondence between the palace and provincial authorities seem to indicate that, by the late 1930s, it was the norm to equip the royal pavilions set up for state or religious celebrations with loudspeakers as they were being wired for electric lights.⁶⁷ Radio broadcasts of the King's Speeches and other royal events, including live Quran recitations from Abdin Palace during religious holidays, were critical components of King Faruq's propaganda apparatus. Faruq's radio exposure complemented the wall-to-wall coverage the young king was already receiving in the Egyptian press. Throughout the succession period and during the first few years of the boy king's reign, the illustrated periodicals in particular, printed hundreds of photographs covering Faruq's every move. Radio broadcasts and their public dissemination through loudspeakers now added a sonic layer to the state's ongoing attempt at legitimating itself to the Egyptian people.

FIGURE 6.8. Loudspeakers in Cairo's Streets.
SOURCE: *al-Radiu al-Misri*, May 30, 1936.

FIGURE 6.9. Nahhas Pasha Reading the King's Speech in Parliament.
SOURCE: *al-Radiu al-Misri*, May 30, 1936.

Conclusion

Royal authority and prestige had its own soundtrack, announcing its legitimizing presence to its subjects whenever the opportunity arose. State cannons and guns were deployed to sound out both state authority and the state's "sponsorship" of important religious festivities. Military parades with marching brass bands playing music, including the "Royal Anthem," added to the din of state power. The advent of state-controlled radio in the summer of 1934, gave the state yet another important sound-making and mediating tool that it used in earnest. Finally, loudspeakers and public address systems were used for live public events and celebrations, and often they were an extension and amplification of the state's radio broadcasts.

These state-produced sounds worked hand in hand with carefully choreographed spectacles and a host of visual representations of the ruling regime's wealth and power. The elaborately organized military parades, displaying Egypt's crack troops and military weapons and machines, visually exhibited the soldiers in their shiny uniforms marching in line, yet also sounding out their footsteps in sonic unison as they stepped to the beat of the brass band's drummers. Not surprisingly, this type of state "noise" was readily deployed and legitimated, even as popular celebrations and other street sounds were decried as inappropriate and offensive, especially when they were associated with the Egyptian masses and public disorder. This was as much about silencing alternative and independent working-class voices as it was about projecting select sounds into public spaces and airways.

The merging of official Islam with state authority was especially effective. Lavishly well-lit and decorated royal pavilions set up for the Mawlid al-Nabi, Ramadan Iftar, Eid, and other celebratory events were de facto mobile palaces, visually and sonically displaying royal authority and prestige. Gun salutes for the king or his representative as he arrived or departed, and fireworks shows at night were the norm at these events. If the king was not there, his voice was perhaps on the radio and amplified through loudspeakers. Although more often, the speakers projected music. The visual and auditory projection of power in these pavilions was softened somewhat with royal displays of piety and/or benevolence, either in the form of providing the people with Quran recitation or by feeding the poor. Thus in many ways, it was a multisensory indulgent treat meant to impress the masses. Appropriation of popular religion—be it through the official celebration of the Mawlid al-Nabi or a state Ramadan celebration focusing on feeding the poor—served two primary purposes: imbuing the king, and by extension the state, with pious authority; and providing official alternatives to the (unfairly maligned) celebratory manifestations of popular Islam. By the 1930s, state anxiety over the growing popularity of the Muslim Brotherhood and its growing influence on the urban poor added yet another motivation for enhancing the state's Islamic and charitable appeal.[68]

The loud government-sponsored street parades and state sounded celebrations discussed in this chapter may have impressed some people, yet they were also most certainly a regular inconvenience to many, as they blocked roads and routinely obstructed traffic. When such disruptions take place regularly, they may have an effect that is the opposite of projecting power: revealing the insecurities of a fragile state in desperate search of legitimacy. It is important as well to keep in mind that we cannot entirely account for how everyday Egyptians, in all their diversity, perceived or internalized all this state noise. Many could have simply tuned it out. Also, although this chapter considers mostly state authority and the sonic projection of that power, we should remain aware that the Egyptian streets and everyday people "spoke" back, using some of the same tools first deployed by the state. As I examined in the last chapter, members of the same military brass bands that played European and official state music in their official functions, spawned a host of groups playing local and popular street music, including the iconic Hasaballah street bands. And the same loudspeakers deployed by the state in the 1930s would soon be used by religious institutions, mosques, and even by ordinary people in their weddings, funerals, *mawalid*, and circumcision celebrations.

Conclusion
Class Distinction and Remembering Lost Sounds

THERE ARE MANY MORE HISTORICAL perspectives to be discovered if we are open to considering sound as a serious avenue for understanding the past. Each historic time and locale has its varied natural, animal, and human sounds, which play an important role in defining the place to its inhabitants. As documented in this book, these sounds and noises have an assortment of economic, environmental, social, and cultural implications, which are vital for a more comprehensive understanding of the past. Studying these sounds is especially important in examining societies at the turn of the twentieth century, as that is arguably the most rapidly changing period of technological and demographic transition in human history. In the same way that all of our five senses are relevant to our daily understanding of the world around us, they should also be vital to our historical understanding of the past. Interpreting how peoples of the past sensorially experienced their world makes possible a richer, more comprehensive grasp of historical events. Simply put, a sensorially grounded historical narrative is an embodied history that is connected to everyday people and to everyday lives.[1]

As we've explored throughout this book, the transformation of Egypt's street environment, from electric lighting to automobile traffic and the growth of modern mass-transportation networks, greatly impacted daily life. The introduction of new technologies, from electrification to the internal combustion engine, dramatically changed how the streets functioned, felt, and sounded. Instead of examining the ruling elite, colonial and Egyptian administrators, nationalist leaders, and luminary writers and journalists, this book's principal focus has been on ordinary people: pedestrians, commuters, beggars, street vendors and entertainers, and the multitude of participants, listeners, and observers in street weddings, funerals, parades, and festivals. *Street Sounds* has also revealed a great many uncertainties, debates, tensions, and accommodations connected to a host of cultural and social issues at the intersection of modernity, class formation, and state power. By limiting my analysis almost entirely to the streets and public squares, I was able to get closer to the realities on the ground for a more embodied investigation of how ordinary Egyptians dealt with these dramatic transformations.

Perhaps the most drastically audible and visible of these technological changes was the impact that electricity had on everyday life, transforming how Egyptians were transported, entertained, and informed. The microphone along with electrical amplification was especially transformative, forever changing how music, speeches, sports events, and films were heard. At first, loudspeakers were used mostly by the Egyptian state for its own informational and propaganda purposes. Yet by the 1940s, they were also used by ordinary people for a variety of other mundane reasons and occasions, including weddings and funerals. Most iconic of course was the gradual use of loudspeakers for the call to prayer, starting in the mid-1940s. Right after the end of World War II, some of the major congregational mosques in Egypt, including the al-Azhar Mosque, wired their minarets with loudspeaker systems. Within a decade, this would forever change the Egyptian soundscape, as the call to prayer became much louder and sounded dramatically different when mediated and amplified through a loudspeaker. It is striking, for example, that most Egyptians today have never heard an unamplified or unmediated call to prayer.

Although the original plans for creating large city squares, widening the roads, and lighting those roads with electric lights were meant to "modernize" and make the streets and alleyways accessible to the ordering powers of state authorities, they equally allowed everyday people from street hawkers to commuters to use these new spaces and technologies for their own purposes and their personal benefit. As is often the case, the same tools used to bring about modernity were counter-used for purposes that were not always in line with how these tools were intended or expected to be used in the first place. The same wide streets and city squares that were meant to allow wheeled vehicles, speed up transportation, and facilitate the penetration of state authorities into even the small alleyways of the city in order to control, tax, and vaccinate, also allowed street merchants and the itinerant poor to make a living throughout the city, including in the newer, upscale neighborhoods. Tramways, in particular, crisscrossed the city and connected the urban slums with the upscale suburbs, creating social anxieties among the elite. Also, although tramways were meant strictly for mass transportation, they were effectively used for commerce by street hawkers and even provided a perfect moving platform for pickpockets to operate. More scandalous to turn-of-the-twentieth-century conservative elites, tramways also created an unintended space for socialization between classes and genders.

Silencing Public "Noise" and Class Distinction

Public noise was usually perceived as dangerously chaotic and socially disruptive, or vulgarized as culturally backward, except when it was produced by the state or when it was invoked as part of a socially broad nationalist mobilization. With few exceptions, especially when dealing with discussions about modernity, the Egyptian media perpetrated a silencing campaign targeting the urban masses. The politics of noise in Egypt, as elsewhere, has a distinctly classist hierarchy, categorizing noise and silence along an arbitrary civilizing continuum, and branding most sounds emanating from the lower classes as vulgar. A critical component of the modernist discourses and the cultural silencing encountered in the Egyptian press was that they reflected the insecurities of the growing, upwardly mobile middle classes attempting to distinguish themselves from the Egyptian masses. Sensory markers were typically used to denote those class distinctions, creating a de facto sensory hierarchy differentiating between the varying levels of society. This was certainly affected and boosted by discourses of western modernity. This middle-class-driven "modernity project" was portrayed in the Egyptian media through simplistic binary distinctions between the ideal, clean, demure, educated middle-class citizens and the loud, backward, diseased, and gullibly superstitious urban masses and peasants. In other words, neutral and pleasant smells, foods, dress, sounds, music, color, tastes, and behavior belonged to the "modern" middle classes and elites; loudness, vulgarity, ignorance, uncleanliness, bad odors, tastelessness, superstition and disease were associated with the vast majority of the Egyptian population. These classist and outright dehumanizing discourses were rationalized somewhat by giving these self-declared modernizers a clear directive to educate and uplift their fellow citizens to an "acceptable" level of civilization.

This systemic, top-down silencing marginalized almost every aspect of traditional Egyptian society, from popular religious celebrations of *mawalid* (saint's day festivals) to traditional healing exorcisms (*zars*). Class anxieties manifested themselves even in relation to popular public expressions of joy, during traditional music-infused celebrations and ululations at weddings, births, and circumcisions. As I examined in Chapter 5, even somber chants as well as shrieks and wails at deaths and funerals were deemed by both modernist-inspired Muslim and Coptic Christian orthodoxies as not only vulgar and immoral but syncretically impure and infected with ancient Egyptian paganism. Most of the attacks on these popular traditions and rituals focused on a sensory critique of the loudness, musicality, and general embodiment and gender mixing at these events. More broadly, even the mundane everyday lower- and working-class "noise" was vulgarized and deemed inferior to the supposedly more "modern" restrained sounds of the upwardly mobile middle classes. The sounds of beggars, street vendors, entertainers, and a host of itinerant poor were especially maligned, and the state was mobilized to regulate and silence the streets.

However, despite state suppression and the incessant elite and middle-class discourses characterizing the backwardness of Egypt's subaltern men and women, for the most part, ordinary Egyptians resiliently and actively resisted or deflected these attacks. Street hawkers, in particular, frequently petitioned the state to redress their grievances, while countless other men and women adapted to all the rapid structural and technological changes by appropriating the very streets and technologies as their own. Ordinary pedestrians, beggars, commuters, paying and non-paying tram passengers, street vendors, and even pickpockets used and "misused" the electric trams, loudspeakers, radio sets, buses, streets, and sidewalks for their own purposes. The watchful eye of the state and the proverbial panopticon was certainly put into use to silence and regulate, yet in the long term, this was largely ineffective. The appropriation of the city's modern transportation systems and infrastructure by ordinary people effectively allowed them to lay claim to modernity and ownership of the streets on their own terms.

Sensory Marginalization between Classism and Nationalism

The vulgarization of Egypt's working class and the "noises" they produce continues until this day. This is especially true when it comes to street hawkers and beggars. They are still very much oppressed by the state and discursively marginalized by the Egyptian upper-middle classes and elites. Often they are portrayed as a mere nuisance, making middle-class life inconvenient by obstructing traffic and by disturbing the peace with their loud calls. When mundane everyday street problems like noise pollution, traffic, or road cleanliness are discussed in the press, or when broader issues of modernity, "progress," or even family planning and overpopulation are invoked, class differentiation and scapegoating of the masses invariably takes place. As we have examined throughout the book, but especially in Chapter 2, this marginalizing discourse is inevitably sensory, embodied, and at times dehumanizing.

However, this marginalization of the Egyptian peasants and the urban working classes is almost completely forgotten whenever Egyptian nationalism is stirred up or flagged in the press. As Selim, Schielke, Gasper, and Ryzova have shown in their examinations of Egyptian novels, autobiographies, and the press, there are deeply rooted contradictions in the portrayals of peasants and the working classes, depending on whether the goal of the author is to emphasize modernity or fire up nationalism.[2] Under the rubric of nationalism, especially in times of national crisis, Egyptian of all classes are supposed to be "equal," and at these moments especially, the middle classes and the cultural elite seek the opposite of differentiation and distinction as they attempt to validate and authenticate their own Egyptianness to some idealized national type. At these moments of national invocation and solidarity, Egyptian peasants and working-class men and women are no longer considered loud, ignorant, tasteless, dirty, and superstitious; overnight they become symbols of resilience, resourcefulness, patience, moderation, and loyalty, representing salt-of-the-earth national authenticity. As Ryzova has shown in her analysis of multiple Egyptian autobiographies, some of these authors will offer contradictory representations of the Egyptian countryside, or of the traditional city quarters, just pages apart.[3]

These apparent contradictions between middle-class distinctions and national authenticity are exemplified in the very words used to demean the Egyptian masses.[4] Throughout the twentieth century, the term *baladi*, which in the Egyptian vernacular literally means "local" or "native," was used as an insult to vulgarize the Egyptian working class as lacking in taste and cultural sophistication. *Baladi* is still used today, though at the turn of the twenty-first century, labeling an Egyptian man or women as *bi'a*, which roughly means "an inhabitant from the local milieu," was also used in the same derogatory manner.[5] These contradictions are consistent with most of the sensory analysis in this book, and help to explain the incongruities and fissures between class identity and national solidarity. In fact, a sensory reading of these writings emphasizes these contradictions even more, as a class-based sensory hierarchy was almost always highlighted whenever modernity or class distinction was invoked.[6]

To be sure though, the classist and the nationalist arguments are not as contradictory as they may first seem. In order to see that, however, we must factor temporality into our analysis. On the one hand, often the most negative descriptions of the peasants and the working classes are describing present-day realities as perceived by the middle classes. Usually these negative portrayals are inspired by and describe current problems, as they are happening, from overpopulation, to traffic and traffic noise, poverty, street crime, the accumulation of garbage in the streets, and other press-driven crises of the day. On the other hand, descriptions of how Egyptians existed in the immediate or even the most distant past, are colored with romanticized nostalgia for an imagined simpler and happier time, when the "authenticating" positive qualities imagined to be "inherent in all Egyptians" were actualized. The same nationalist imagination also projected a more positive future, when, to be sure, the educated middle classes will be the instructive agents of change, uplifting the "unwashed, smelly, raucously-loud, overtly sexualized and superstitious urban and rural masses" and transforming them into "clean, quiet, sober minded, restrained, and rational citizens." Dystopian as this silent and colorless vision might sound, to an eager nationalist this flavorlessly banal future exemplifies progress.

Nostalgia for "Lost" Street Sounds, Nationalism, and Media Celebrations of Everyday Life

> In Salama Street [in Cairo's Sayyida Zaynab district] there were year-round colorful scenes regularly taking place, making it feel as if we were all participants in an entertaining novella.... There was the Fidelity of the Nile [Waffaa' al-Nil] Parade Festival ... and the nightly *misaharati* [Ramadan street crier] making his rounds during the month of Ramadan.... There were the wedding celebration parades that often included the *niqrizan* drum dance.[7] We also witnessed the circumcision celebratory processions [*zaffit al-mitahir* (transporting the children)] on horse-drawn carriages, and the sounds of the beggars who called for alms to feed their newborn children. All these are rich and colorful scenes illustrating the degree to which the lives of the people of Cairo were filled with a beautiful art. The art of

> music and dance are intertwined with, and exemplify the
> inherited beliefs and traditions of everyday people.
>
> Fathi Radwan, *Khati al-Ataba*[8]

Fathi Radwan's childhood autobiography exemplifies a nostalgic genre of writing that romanticizes the past by highlighting a favorable sensory recollection of childhood memories. The very title of Radwan's autobiography—*Khati al-Ataba: Hayyat Tifl Misri* (Pass the threshold: The life of an Egyptian child)—sonically recalls a traditional Egyptian nursery rhyme. "*Tata-tata khati al-'ataba*," which roughly translates to "step by step . . . you will pass the threshold," is still sung by some Egyptian mothers and fathers to encourage their toddlers to walk.[9] As depicted in the passage from his book, many of Radwan's memories relate to sensory recollections of the street and its countless vendors and entertainers. His autobiography is full of vivid descriptions of monkey street performers (*qiradatis*) beating their drums, puppeteers with their wheeled mobile stages, street magicians, picture peepshows (*sunduq al-dunia*), and multiple *zaffah* processions with music and belly dancers.[10]

Throughout his book, Radwan rarely passes judgment on many of the older traditions that often get ridiculed by his contemporaries, offering one of the most sympathetic nonfictional accounts of traditional Egyptian society of his generation. As mentioned earlier, Radwan had a long political public life, from the early 1930s as a co-founder of the oppositional ultranationalist Young Egypt Party to his later, more conformist career as Nasser's Minister of National Guidance. Part of the reason for Radwan's sentimental and generally positive view of tradition was that he wrote and published his book—which depicts his childhood from the mid-1910s to the mid-1920s—a full half century after that time. Nostalgia for the distant and disappearing sights, smells, and sounds of his childhood, coupled with his fervent nationalism, contributed no doubt to his idealistic depiction of the past. Indeed, most of the sights and sounds that Radwan describes in his book were already gone or about to disappear for good by the time he was writing about them in the 1970s. With the onset of increasing urbanization and modernization during the twentieth century, most cultures develop a sense of nostalgia for sounds of the past. Mundane sounds that were ignored, taken for granted, or even complained about and vulgarized, are rightly or wrongly romanticized, and remembered fondly after they have been lost.

Sound studies scholar R. Murray Schafer appropriately asks: "Where are the museums for disappearing sounds? Even the most ordinary sounds will be affectionately remembered after they disappear."[11] Throughout the twentieth century, Egypt's entertainment industry somewhat served that role of remembering, and in an imperfect and heavily mediated form, memorialized some of the disappearing sounds and experiences of the past.[12] As I examined in Chapter 1, for example, by the first decade of the twentieth century it was well understood that the profession of the *saqa*, the water seller, was becoming obsolete. Gradually, though consistently, the sounds of the water sellers' calls disappeared from the urban soundscape. As with many of the changing social and cultural realities of early-twentieth-century Egypt, the ordeals of the Egyptian water sellers were captured and memorialized—though with typical lighthearted comedic flair—by the musical compositions of Sayyid Darwish and the lyrics of Badi' Khayri. Sayyid Darwish was one of the earliest composers/musicians to infuse his music with a deeply reflective documentation of street life, celebrating in the process multiple Egyptian voices and sounds. In the case of his 1918 "Lahn al-Saqayyin" (Water sellers' song), not only does the song commemorate the way of life of the *saqayin*, but in an imperfect way it preserves at least the memory, if not the pitch and tone, of their calls.[13]

"Lahn al-Saqayyin" was also called "Yi'awad Allah" (God will compensate), which was one of the traditional street cries of the water sellers. It was written after extensive research, with both Darwish and Khayri sitting for hours in a coffee shop near the water sellers' quarter (*harat al-saqayin*) in Cairo. They spoke with and interviewed the water sellers and, more importantly, listened to and copied the tone and pitch of their calls.[14] In the song, after repeating the chorus "God will compensate," the singing water sellers begin to list their complaints and grievances, while acknowledging the inevitable obsolescence of their profession.[15] Despite its farcical nature, the song has an undertone of collective nostalgia as it documents and acknowledges a disappearing way of life. The song resonated with many of its listeners not just because of its nationalist message, but because, through the lens of the water sellers, it captured the reality (which was uncomfortable to some) that Egypt during the 1920s and 1930s was a society very much in transition.

Much of Sayyid Darwish's music, along with words of his lyricists Badi' Khayri, Yunis al-Qadi, and Bayram al-Tunsi, celebrated the sounds of Egyptian everyday life. The songs of these musicians were especially adept at capturing, mediating, and amplifying the sounds and emotions of the Egyptian streets, while also invoking nationalism and national unity. From the end of World War I until the premature death of Darwish in 1923, dozens of songs were recorded that celebrated and recounted the daily hardships of Egypt's peasants and the working class. These songs covered the plight of cargo porters carrying heavy loads, workers toiling in the Egyptian factories, peasants working in the fields and even street sweepers, beggars, and fortune tellers were positively represented as hard-working Egyptians dealing with all of the social and economic changes exacted by modernity and striving to make an honest living.[16]

Perhaps the best and most widely known of this genre of nostalgic and celebratory music is the puppet play operetta *al-Layla al-Kabira* (The great night). Al-Layla al-Kabira is the name given to the last night of a *mawlid*, a saint's day festival; and as the name suggests, it typically is the climax, the loudest and most celebratory night of the festival. The vernacular lyrics of this state-sponsored musical were written by Salah Jahin, and the music was composed by Sayyid Mikawi.[17] It was first performed in the Egyptian National Puppet Theatre in 1960, and shortly after, it was broadcast on Egyptian National Television. For almost sixty years, the thirty-five-minute musical was repeatedly performed and re-broadcast on Egyptian television and radio. Subsequently, multiple audio versions of the musical, some as short as ten minutes, were re-recorded and distributed by many artists. *Al-Layla al-Kabira* is still universally known and loved, and Egyptians of all classes and ages know some of its lyrics by heart and will often sing along when the music is played.[18]

In a masterful scene-by-scene whiplash of musical and lyrical improvisation, *al-Layla al-Kabira* recounts—albeit in a heavily mediated form—some of the sounds and musicality of a typical *mawlid*. Each of the short, fifteen- to thirty-second vignettes covers a particular aspect of the festival, as if experienced from a pedestrian's perspective. The majority of the scenes musically recreate the calls of various food sellers, street hawkers, shooting gallery and carnival strength-game operators, fairground wrestlers, and even a circus lion tamer. Singing cameo appearances are made by a belly dancer, an Upper Egyptian singer, a photographer advertising his (6 cm by 9 cm) photos, coffee shop owners, waiters and their guests, and the participants in a circumcision procession. The various vernacular accents of Egypt are represented, the most notable of which is the Upper Egyptian (*sa'idi*) accent of an *'umda* (village mayor), who comically gets lost in the maze of the city.[19] Although the operetta is a heavily mediated representation, and is much tamer than an actual Egyptian *mawlid*, it retains and amplifies a carnival-like, light-humored, and celebratory feel.

Nevertheless, there are some obvious contradictions between the actual *mawalid*—which for more than a century have been neglected by the state and are often critiqued and vulgarized by intellectuals, modernists, and Islamists— and their fictionalized representation in Egyptian media as a glorified, if ossified, folkloric ideal exemplifying national authenticity. In fact, when Jahin wrote *al-Layla al-Kabira* in 1960, Egyptian *mawalid* were already in decline and were no longer attracting a large cross-section of the Egyptian population. Today, most of the major Egyptian *mawalid* are still celebrated, though on a smaller scale than in the past, while dozens of smaller *mawalid* have altogether disappeared. Unlike at the turn of the twentieth century, today there are many Egyptians (mostly middle- and upper-class urbanites) who have never attended a *mawlid*, and for them *al-Layla al-Kabira* is the only *mawlid* they know.[20] In other words, while the actual *mawalid* have been "either sanitized or removed from the public imagery of the nation," virtual media representations are taking their place, but only after they have been tamed, and transformed into an integral part of the "folkloric culture of the modern state."[21] It would seem that—just as we have seen with street hawkers, peasants, and the urban working poor—actual flesh and blood *mawalid* participants are at best silenced or rendered invisible by the state, yet their fetishized and sanitized media representations are loudly portrayed as part of an imagined and idealized national collective.

. . .

As the very fabric of the streets modernized and changed and more and more of the familiar keynote sounds declined and then altogether disappeared, the "memories" or representations of these sounds—even the sounds of street hawkers—have become nostalgically ingrained and reimagined in songs, movies, miniseries, and more recently, in an increasing number of Facebook pages and YouTube channels. Ironically, the same peasants and working-class Egyptians who are often sensorially marginalized and vulgarized in everyday life are transformed in some of their media representations into an essential part of the nation and imbued with national authenticity. It seems as though when nostalgia and/or nationalism are invoked, an egalitarian generosity tends to be overemphasized in the mass media, and for a time at least, brings the nation with all its class, racial, and religious diversity under the same banner.

In many ways, the purpose of this book was not to recover the lost sounds and soundscapes of the past. That would not only be futile, as it is impossible to do, but it also would not help us understand the historical significance of these sounds. As I have shown in these pages, history and historical sources are rich with sonic accounts. The question is, can historians find evidence of, and take account of, these changing sounds to determine their sociocultural and perhaps even their economic and political significance? This book, and the growing field of sound studies, emphatically answers this question in the affirmative. By heeding the voices of ordinary people and by documenting how they physically occupied, used, and misused the streets, it has been possible to uncover some of their agency and, at times, even their successes in resisting the increasing panoptic powers of the state. Ordinary people spoke out, sang, and sounded out their thoughts and emotions constantly and in a variety of ways. In other words, by listening in to modernity's changing soundscapes and the sounds of everyday people engaging and sometimes clashing with these transformations, we are able to grasp various street-level perspectives on class formation, state power, and the impact of modernity on everyday life.

Notes

Preface

1. Even seemingly banal sounds can reveal a great deal about a place. For example, analyzing and historicizing the ostensibly simple call of *bikyya* reveals that the colloquial Egyptian term *robabikya* is derived from the Italian *roba vecchia*, which means "old stuff." There was a large working-class Italian community in Egypt from the mid-nineteenth to the mid-twentieth century, and it is likely that some of them took on the trade of junk removal. Their calls of *roba vecchia*, or simply *vecchia*, became synonymous with the meaning "junk," and entered the Egyptian vernacular lexicon sometime around the turn of the twentieth century. In any case, the name for this profession and its unique calls continues to live on, albeit in the Egyptianized form *robabikya*.

Introduction

Parts of this introduction were previously published elsewhere. See Ziad Fahmy, "An Earwitness to History: Street Hawkers and Their Calls in Early Twentieth Century Egypt," Roundtable, *International Journal of Middle East Studies* 48, no. 1 (2016): 129–134; and Ziad Fahmy, "Coming to Our Senses: Historicizing Sound and Noise in the Middle East," *History Compass* 11, no. 4 (April 2013): 305–315.

1. *Al-Radiu al-Misri*, February 29, 1936.

2. U.S. Department of Commerce, Commerce Reports, *Agricultural and Industrial Exhibition, Cairo* (June 26, 1936), 493. This figure likely does not take into account repeat visitors to the exhibition.

3. J. W. Taylor, *Egypt: Economic and Commercial Conditions in Egypt*, Overseas Economic Surveys (London: His Majesty's Stationery Office, 1948), 54.

4. *Al-Radiu al-Misri*, February 15, 1936; *al-Musawwar*, February 25, 1949. The exhibition was primarily sponsored by the Royal Agricultural Society of Egypt. Owing to World War II, the next exhibition would not be held until 1949.

5. Timothy Mitchell, *Colonising Egypt* (Berkeley: University of California Press, 1991), 1–31.

6. For an excellent history of the architectural and visual representations of Egypt's twentieth-century exhibitions, see Mohamed Elshahed, "Egypt Here and There: The

Architectures and Images of National Exhibitions and Pavilions, 1926–1964," *Annales Islamologiques* 50 (2016): 107–143.

7. Mitchell, *Colonising Egypt*, 1–31.

8. *Al-Radiu al-Misri*, February 15, 1936.

9. *Al-Radiu al-Misri*, February 15, 1936; February 29, 1936.

10. This book focuses primarily on the street sounds of urban areas. However, similar changes, albeit on a much smaller scale, were simultaneously happening in rural areas. Water pumps for irrigation, trains and train whistles, automobile and bus traffic, and eventually electricity and loudspeakers made a sonic impact on the Egyptian countryside.

11. For an excellent examination of some of these technological changes, from trains to telephones and telegraphs, see On Barak, *On Time: Technology and Temporality in Modern Egypt* (Berkeley: University of California Press, 2013).

12. For more on the transitional nature of newer industrial and modern sounds as gradually added to "older sounds," see Mark M. Smith, "The Garden in the Machine: Listening to Early American Industrialization," in *The Oxford Handbook of Sound Studies*, ed. Trevor Pinch and Karin Bijsterveld (Oxford: Oxford University Press, 2011), 39–57.

13. Electrification typically spread gradually from well-to-do districts in the early twentieth century to more popular urban and even rural areas from the 1920s to the 1950s. See Chapters 3 and 4 for more detail on the electrification of Cairo.

14. David Howes, *The Sixth Sense Reader* (London: Berg, 2009), 35; Michael Bull et al., "Introducing Sensory Studies," *The Senses and Society* 1 (2006): 5–7.

15. Mark M. Smith, "Introduction: Onward to Audible Pasts," in *Hearing History: A Reader*, ed. Mark M. Smith (Athens: University of Georgia Press, 2004), ix.

16. See "Roundtable: Bringing Sound into Middle East Studies," *International Journal of Middle East Studies* 48, no.1 (2016): 113–155; this *IJMES* roundtable was led by Andrea Stanton and Carole Woodall, with contributions from Deborah Kapchan, Michael O'Toole, Lauren Osborne, Ziad Fahmy, Camron Amin, and Ida Meftahi. For other works that take sound into account, see, for example, Ziad Fahmy, *Ordinary Egyptians: Creating the Modern Nation through Popular Culture* (Stanford: Stanford University Press, 2011); Andrea Stanton, *"This Is Jerusalem Calling": State Radio in Mandate Palestine* (Austin: University of Texas Press, 2014); Carole Woodall, "Sensing the City: Sound, Movement, and the Night in 1920s Istanbul" (PhD diss., New York University, 2008); Adam Mestyan, "Sound, Military Music, and Opera in Egypt during the Rule of Mehmet Ali Pasha (r. 1805–1848)," in *Ottoman Empire and European Theatre*, vol. 2 of *The Time of Joseph Haydn—From Sultan Mahmud I to Mahmud II (r. 1730–1839)*, ed. Michael Huttler and Hans Ernst Weidinger (Vienna: Hollitzer Wissenschaftsverlag, 2014), 631–656; and Adam Mestyan, "Upgrade? Power and Sound during Ramadan and 'Id al-Fitr in the Nineteenth-Century Ottoman Arab Provinces," *Comparative Studies of South Asia, Africa and the Middle East* 37, no. 2 (2017): 262–279. For a more detailed argument about the need to incorporate sound in historical research on the Middle East, see Fahmy, "Coming to our Senses."

17. For a more detailed examination of some of the latest anthropological and media studies work on the modern Arab World, see Mohamed Zayani, "Toward a Cultural

Anthropology of Arab Media: Reflections on the Codification of Everyday Life," *History and Anthropology* 22, no. 1 (2012): 37–56; and Walter Armbrust, "Audiovisual Media and History of the Middle East," in *History and Historiographies of the Modern Middle East*, ed. Amy Singer and Israel Gershoni (Seattle: University of Washington Press, 2006), 288–312; Deborah Kapchan, ed., *Theorizing Sound Writing* (Middletown, CT: Wesleyan University Press, 2017).

18. Charles Hirschkind, *The Ethical Soundscape: Cassette Sermons and Islamic Counterpublics* (New York: Columbia University Press, 2006).

19. Jonathan Sterne, *The Audible Past: Cultural Origins of Sound Reproduction* (Durham, NC: Duke University Press, 2003), 3.

20. My examination of class distinctions using noise and sound politics emulates the success of sound studies scholars who have examined American racism and race formation using the racial politics of sound. Much of Mark M. Smith's writing is relevant here and, more recently, the work of Jennifer Lynn Stoever and Nina Sun Eidsheim. Jennifer Lynn Stoever, T*he Sonic Color Line: Race and the Cultural Politics of Listening* (New York: New York University Press, 2016); Nina Sun Eidsheim, *The Race of Sound: Listening, Timbre, and Vocality in African American Music* (Durham, NC: Duke University Press, 2019).

21. For the most comprehensive examination of Egyptian twentieth-century middle-class formation, see Lucie Ryzova, *The Age of the Efendiyya: Passages to Modernity in National Colonial Egypt* (Oxford: Oxford University Press, 2014).

22. The Egyptian press was full of editorials calling for regulating the calls and sounds of street vendors, marginalizing the traditional and festive sounds of weddings and saint's day celebrations and even vulgarizing the traditional wails sounded out during deaths and funerals.

23. For a look at the evolving definition of noise within a European context, see R. Murray Schafer, *The Soundscape: Our Sonic Environment and the Tuning of the World* (Rochester, NY: Destiny Books, 1994 [1977]), 182–183; and Hillel Schwartz, *Making Noise: From Babel to the Big Bang and Beyond* (New York: Zone Books, 2016), 1–30.

24. For an excellent examination of working-class Egyptian men and women during the first half of the twentieth century, see Hanan Hammad, *Industrial Sexuality: Gender, Urbanization, and Social Transformation in Egypt* (Austin: University of Texas Press, 2016).

25. Leigh Eric Schmidt, *Hearing Things: Religion, Illusion, and the American Enlightenment* (Cambridge, MA: Harvard University Press, 2002), 22.

26. In the mid-twentieth century, ocularcentrism was somewhat popularized by Walter Ong and Marshall McLuhan's "great divide" theory. In his *Understanding Media*, for example, McLuhan describes "the inability of oral and intuitive oriental culture to meet with the rational, visual European patterns of experience." In *The Gutenberg Galaxy*, this line of thinking takes on an extreme binary: "there can be no greater contradiction or clash in human cultures than that between those representing the eye and the ear." Marshall McLuhan, *Understanding Media: The Extensions of Man* (New York: McGraw-Hill, 1964), 5; Marshall McLuhan, *The Gutenberg Galaxy: The Making of Typographic Man* (Toronto: University of Toronto Press: 1962), 68. For an excellent critique and analysis of Ong and McLuhan's theories, see Schmidt, *Hearing Things*, 1–22.

27. Schmidt, *Hearing Things*.

28. For example, see Alain Corbin, *Village Bells: Sound and Meaning in the Nineteenth-Century French Countryside* (New York: Columbia University Press, 1998); Alain Corbin, *The Foul and the Fragrant: Odor and the French Social Imagination* (Cambridge, MA: Harvard University Press, 1988); Alain Corbin, *Time, Desire and Horror: Towards a History of the Senses* (Cambridge: Polity Press, 1995); Schmidt, *Hearing Things*; Mark M. Smith, *Sensing the Past: Seeing, Hearing, Smelling, Tasting, and Touching in History* (Berkeley: University of California Press, 2008); Hirschkind, *Ethical Soundscape*; and Sterne, *Audible Past*.

29. Smith, *Sensing the Past*, 48.

30. It is perhaps best to leave such broad pronouncements about the sensory objectivity of eyes versus ears to neuroscientists and cognitive philosophers.

31. Bryan Reynolds and Joseph Fitzpatrick, "The Transversality of Michel de Certeau: Foucault's Panoptic Discourse and the Cartographic Impulse," *Diacritics* 29, no. 3 (1999): 63–80.

32. Michel Foucault, *The Birth of the Clinic: An Archaeology of Medical Perception* (New York: Random House Vintage Books, 1973), 39.

33. It is worth noting that Bentham's model for a panopticon prison was itself never fully implemented and built, though it inspired prison builders to use some of its components in their designs.

34. For instance, neither the Egyptian Lancaster schools nor the model villages discussed by Timothy Mitchell in *Colonising Egypt* were successful in the long term. Their failure reminds us that there is a significant gap between theory and practice, between plans and their execution and reception in the streets.

35. Henri Lefebvre, *The Production of Space* (Malden: Blackwell, 1991), 4.

36. Lefebvre, *Production of Space*, 62.

37. Lefebvre, *Production of Space*, 4.

38. Michel de Certeau, *The Practice of Everyday Life* (Berkeley: University of California Press, 2011), 117.

39. De Certeau, *Practice of Everyday Life*, 95.

40. Lefebvre, *Production of Space*, 12.

41. Steven Connor, "Edison's Teeth: Touching Hearing," in *Hearing Cultures: Essays on Sound, Listening and Modernity*, ed. Veit Erlmann (New York: Berg, 2004), 153.

42. Yi-Fu Tuan, *Space and Place: The Perspective of Experience* (Minneapolis: Minnesota University Press, 2014 [1977]), 11.

43. Connor, "Edison's Teeth," 156.

44. Studies that have successfully used a sensory approach include Peter Charles Hoffer, *Sensory Worlds in Early America* (Baltimore: John Hopkins University Press, 2003); Caroline A. Jones, *Eyesight Alone: Clement Greenberg's Modernism and the Bureaucratization of the Senses* (Chicago: University of Chicago Press, 2005); Roger Horowitz, *Putting Meat on the American Table: Taste, Technology, Transformation* (Baltimore, Johns Hopkins University Press, 2006); Mark M. Smith, *How Race Is Made: Slavery, Segregation,*

and the Senses (Chapel Hill: University of North Carolina Press, 2006); Paul Rodaway, *Sensuous Geographies: Body, Sense and Place* (London: Routledge, 1994); and Corbin, *The Foul and the Fragrant*.

45. Richard C. Rath, "Hearing American History," *Journal of American History* 95, no. 2 (2008): 417.

46. Schafer, *Soundscape*, 8.

47. To paraphrase Bruce Smith, historians must learn to "hear, and not just see, the evidence encoded on pieces of paper." Bruce Smith, "Listening to the Wild Blue Yonder: The Challenge of Acoustic Ecology," in *Hearing Cultures: Essays on Sound, Listening and Modernity*, ed. Veit Erlmann (New York: Berg, 2004), 24.

48. For one of the few articles on olfactory history by a historian of the Middle East, see Khaled Fahmy, "An Olfactory Tale of Two Cities: Cairo in the Nineteenth Century," in *Historians in Cairo: Essays in Honor of George Scanlon*, ed. Jill Edwards (Cairo: American University in Cairo Press, 2002), 155–187. For sensory recollections of Cairo's "Jewish Quarter," see Deborah Starr, "Sensing the City: Representations of Cairo's Harat al-Yahud," *Prooftexts* 26 (2006): 138–162.

49. See, for example, Mark M. Smith, "When Seeing Makes Scents," *American Art* 24 (2010): 12–14.

Part I

1. Fathi Radwan, *Khati al-Ataba: Hayyat Tifl Misri* (Cairo: Dar al-Ma'arif, 1973), 126. For more on the Egyptian author and politician Fathi Radwan, whose *Khati al-Ataba: Hayyat Tifl Misri* (Pass the threshold: The life of an Egyptian child) is cited a number of time in this book, and his political career from the 1930s as a co-founder of the ultranationalist Young Egypt Party to becoming Gamal Abdul Nasser's Minister of National Guidance, see Arthur Goldschmidt Jr., *Biographical Dictionary of Modern Egypt* (London: Lynne Rienner, 2000), 79; Israel Gershoni and James Jankowski, *Redefining the Egyptian Nation, 1930–1945* (New York: Cambridge University Press, 1995), 4, 99; Israel Gershoni and James Jankowski, *Confronting Fascism in Egypt: Dictatorship Versus Democracy in the 1930s* (Stanford: Stanford University Press, 2010), 243–250; Noor Khan, *Egyptian-Indian Nationalist Collaboration and the British Empire* (New York: Palgrave Macmillan, 2011), 118–121; and Joel Gordon, *Nasser's Blessed Movement: Egypt's Free Officers and the July Revolution* (Oxford: Oxford University Press, 1992), 77, 100.

2. Ministry of the Interior Report for the Year 1922, TNA, FO 141/586/4.

Chapter 1

1. Michel de Certeau, *The Practice of Everyday Life* (Berkeley: University of California Press, 2011), 93.

2. R. Murray Schafer, *The Soundscape: Our Sonic Environment and the Tuning of the World* (Rochester, NY: Destiny Books, 1994 [1977]), 8.

3. *Cairo street scenes—outtakes*, Fox Movietone News Story 2-18, 1928 [originally filmed on December 28,1928], http://mirc.sc.edu/islandora/object/usc%3A32904. The

film is available at the University of South Carolina's film archives, and parts of it can be accessed online.

4. For more on 'afiyas, see Ziad Fahmy, *Ordinary Egyptians: Creating the Modern Nation through Popular Culture* (Stanford: Stanford University Press, 2011), 70–71.

5. See *Cairo street scenes—outtakes*, from 8:56 to 9:30.

6. *Nazzaarat al-Dakhiliyya: Qawanin wa-lawa'ih al-Bulis al-Misri* (Cairo: al-Matba'a al-Kubra al-Amiriyya, 1895), 546. Section 1 of the law stipulated the bell requirement. The law was applied on March 12, 1894.

7. *Nazzaarat al-Dakhiliyya*, 510. Section 17 of the law stipulated the whistle requirement. The law was applied on July 26, 1894.

8. *Nazzaarat al-Dakhiliyya*, 436. Section 6 of the law stipulated the requirement to be quiet. The law was applied on January 10, 1891.

9. See *Cairo street scenes—outtakes*, from 0:24 to 0:33.

10. For a theoretical discussion on the importance of time and space, see Doreen Massey, *For Space* (London: Sage, 2015).

11. For a brief history of the spread of carriages and cabbies in Egypt, see John T. Chalcraft, *The Striking Cabbies of Cairo and Other Stories: Crafts and Guilds in Egypt, 1863–1914* (Albany: State University of New York Press, 2005), 59–60.

12. Gabriel Charmes, *Five Months at Cairo and in Lower Egypt* (London: Bentley Press, 1883), 77.

13. Douglas Sladen, *Oriental Cairo: The City of the Arabian Nights* (Philadelphia: Lippincott, 1911), 69.

14. De Certeau, *Practice of Everyday Life*, 97–98.

15. See, for example, Khaled Fahmy, "An Olfactory Tale of Two Cities: Cairo in the Nineteenth Century," in *Historians in Cairo: Essays in Honor of George Scanlon*, ed. Jill Edwards (Cairo: American University in Cairo Press, 2002), 155–187.

16. De Certeau, *Practice of Everyday Life*, 97.

17. De Certeau, *Practice of Everyday Life*, 97.

18. Hillel Schwartz, *Making Noise: From Babel to the Big Bang and Beyond* (New York: Zone Books, 2016), 430.

19. A *ghafir*'s primary role, whether he was employed by the government or business owners, was to guard neighborhoods, storefronts, businesses, and other private and public properties, especially during the night. For government-employed *ghafirs* and some of the guard duties they were assigned, see Ministry of Public Works, *Annual Report for 1921–1922* (Cairo: Government Press, 1925), 85.

20. Charmes, *Five Months at Cairo*, 26.

21. For descriptions of the *ghafir*'s calls, see Charmes, *Five Months at Cairo*, 26; and M. L. Whately, *Child-Life in Egypt* (Philadelphia: American Sunday School Union, 1866), 56.

22. For more on the *effendiyya* and their cultural and clothing habits, see Lucie Ryzova, *The Age of the Efendiyya: Passages to Modernity in National Colonial Egypt* (Oxford: Oxford University Press, 2014).

23. Douglas Sladen, *Queer Things About Egypt* (London: Hurst & Blackett, 1911), 189.

24. For more on the prevalence and function of gold jewelry in Egypt, see Evelyn Early, *Baladi Women of Cairo: Playing with an Egg and a Stone* (London: Lynne Reimer, 1993), 5–6.

25. For a "modernist," early-twentieth-century critique of Egyptian women and the "excessive wearing" of jewelry, see Jirjis Antun, *al-'Insaniyyah wa-al-Tamadun* (Cairo: Matba'at al-Ma'arif, 1912), 90–92.

26. Sladen, *Oriental Cairo*, 68.

27. Sladen, *Oriental Cairo*, 92.

28. Early, *Baladi Women of Cairo*, 5.

29. For more detail on veiling and unveiling, see Leila Ahmed, *Women and Gender in Islam: Historical Roots of a Modern Debate* (New Haven: Yale University Press, 1993); and Leila Ahmed, *A Quiet Revolution: The Veil's Resurgence, from the Middle East to America* (New Haven: Yale University Press, 2012).

30. Whately, *Child-Life in Egypt*, 30.

31. André Raymond, *Cairo: City of History* (Cairo: American University in Cairo Press, 2007), 246.

32. Raymond, *Cairo*, 235, 359.

33. Gabriel Baer, *Egyptian Guilds in Modern Times* (Jerusalem: Israel Oriental Society, 1964), 36–37.

34. Sladen, *Oriental Cairo*, 304.

35. Whately, *Child-Life in Egypt*, 306.

36. Schafer, *Soundscape*, 64–65.

37. Whately, *Child-Life in Egypt*, 32–35. Throughout the decades that Mary Louisa Whately lived in Egypt, she built a variety of schools for girls and, later on, boys in Cairo and elsewhere in Egypt. A missionary and devout Christian, she is revealed by her prose to be biased when it comes to matters of religion, yet she is sympathetic and fair-minded when writing about her Egyptian students, friends, and neighbors. She is also an astute listener and observer of the sights and sounds she witnessed in her everyday life in Egypt.

38. Muhammad 'Umar, *Hadir al-Misriyyin 'Aw Sirr Ta'akhurihum* (Cairo: Matba'at al-Muqtataf, 1902), 227–229.

39. See James Heyworth-Dunne, "A Selection of Cairo's Street Cries (Referring to Vegetables, Fruit, Flowers and Food)," *Bulletin of the School of Oriental Studies* 9 (1938): 351–361.

40. Radwan, *Khati al-Ataba*, 93.

41. Heyworth-Dunne, "A Selection of Cairo's Street Cries," 352–354.

42. Radwan, *Khati al-Ataba*, 92–93.

43. Whately, *Child-Life in Egypt*, 34–35.

44. To watch and listen to this juice seller, see *Cairo street scenes—outtakes*, from 4:32–4:40.

45. Whately, *Child-Life in Egypt*, 32–33.

46. Whately, *Child-Life in Egypt*, 32–33.

47. Constance Fenimore Woolson, *Mentone, Cairo and Corfu* (New York: Harper & Brothers, 1896), 212.

48. Whately, *Child-Life in Egypt*, 34–35.

49. William Lawrence Balls, *Egypt of the Egyptians* (London: Pitman & Sons, 1920), 104. During the first half of the twentieth century, Balls was a botanist in the Khedivial Agricultural Society of Egypt in Cairo and, later, the Egyptian Department of Agriculture.

50. Sladen, *Oriental Cairo*, 66.

51. Egyptian *friska* are derived from Italian *pizzelles*, or wafer cookies, most likely introduced to Alexandrians by that city's large Italian community at the turn of the twentieth century. The term *friska* comes from the Italian *fresca*, meaning "fresh." It is very likely that some of the early *friska* sellers in Alexandria were Italian street hawkers calling out *"pizzelles fresca!"* The Alexandrian words for ice cream (*jilati*), lemonade (*limonata*), and sea urchin (*ritsa*) are also derived from the Italian: *gelato, limonata,* and *riccio di mare*.

52. Whately, *Child-Life in Egypt*, 34–35.

53. Whately, *Child-Life in Egypt*, 29.

Chapter 2

1. Pierre Bourdieu, *Distinction: A Social Critique of the Judgment of Taste* (Cambridge, MA: Harvard University Press, 1984), 7.

2. In the same vein as the vulgarizing attacks by the Egyptian cultural elites on most forms of vernacular popular culture (which I discuss in *Ordinary Egyptians*), these attacks on Egyptian street people take on a distinctly sensory dimension.

3. For an excellent and detailed analysis of the cultural and economic makeup of the effendiyya and the Egyptian middle classes, see Lucie Ryzova, *The Age of the Efendiyya: Passages to Modernity in National Colonial Egypt* (Oxford: Oxford University Press, 2014). For a well-researched study on masculinity and middle-class formation in Egypt, see Wilson Chacko Jacob, *Working Out Egypt: Effendi Masculinity and Subject Formation in Colonial Modernity, 1870–1940* (Durham, NC: Duke University Press, 2011).

4. Bourdieu, *Distinction*, 6–7.

5. Bourdieu, *Distinction*, 77.

6. Muhammad 'Umar, *Hadir al-Misriyyin 'Aw Sirr Ta'akhurihum* (Cairo: Matba'at al-Muqtataf, 1902).

7. Edmond Demolins, *A quoi tient la supériorité des Anglo-Saxons* (Paris: Firmin-Didot, 1898).

8. For more on the Dinshaway incident, see Ziad Fahmy, *Ordinary Egyptians: Creating the Modern Nation through Popular Culture* (Stanford: Stanford University Press, 2011), 92–95.

9. Ahmad Fathi Zaghlul, *Sirr Taqaddum al-Injliz al-Saksuniyyin* (Cairo: Matba'at al-Ma'arif, 1899).

10. Zachary Lockman, "Imagining the Working Class: Culture, Nationalism, and Class Formation in Egypt, 1899–1914." *Poetics Today* 15, no. 2 (Summer 1994): 163.

11. For an excellent synopsis of 'Umar's argument with regard to class, see Lockman, "Imagining the Working Class," 157–191.

12. Lockman, "Imagining the Working Class," 168.

13. For more on many of these discourses, and the tensions between the benevolent and the classist impulses of Egypt's elites, see Mine Ener, *Managing Egypt's Poor and the Politics of Benevolence, 1800–1952* (Princeton: Princeton University Press, 2003).

14. For typical anti-populist attitudes and general fear of the masses mobilizing in the streets, see Mikha'il Sharubim, *Al-Kafi fi Tarikh Misr al-Qadim wa al-Hadith*, vol. 5, bk. 3 (Cairo: Matba'it Dar al-Kutub wa al-Watha'iq al-Qawmiyya bil-Qahira, 2003), 492–500.

15. 'Umar, *Hadir al-Misriyyin 'Aw Sirr Ta'akhurihum*, 227.

16. 'Umar, *Hadir al-Misriyyin 'Aw Sirr Ta'akhurihum*, 229–230.

17. Unsuccessful attempts at regulating, documenting, and taxing them began in the mid-nineteenth century and continue to this day. Indeed, as I personally encountered many times on my research trips to Cairo and Alexandria in 2013 and 2014, police abuse and harassment of street merchants and hawkers very much continues today. The regular cat-and-mouse games between the police and street merchants near Azbakiyya Square in Cairo today are a testament to the continuing tug-of-war between street merchants and the state.

18. "Monsieur Eram's proposed concession to the president of Majlis al-Nuzzar" (in French), DWQ, Majlis al-Wizara' Files, no. 0075-022228 (January 12, 1889).

19. "Explanatory memo on the history of regulating street hawkers" (in Arabic), DWQ, Majlis al-Wizara' Files, no. 0075-055053 (July 1941).

20. "Laws and regulations regarding street hawkers" (in Arabic), DWQ, Majlis al-Wizara' Files, no. 0075-055053 (July 1941).

21. A millime is one-thousandth and a piaster is one-hundredth of an Egyptian pound (LE).

22. "Laws and regulations regarding street hawkers."

23. For an excellent examination of contemporary police relations with itinerant merchants, see Salwa Ismail, *Political Life in Cairo's New Quarters: Encountering the Everyday State* (Minneapolis: University of Minnesota Press, 2006).

24. "Petition to King Faruq by Abd al-Maqsud Khalaf Hasan" (in Arabic), DWQ, Watha'iq 'Abdin Files, no. 0069-010545 (October 16, 1944).

25. "Petition to King Faruq by Abd al-Maqsud Khalaf Hasan."

26. "Petition to King Faruq by Selim Nicola on behalf of a dozen street merchants" (in Arabic), DWQ, Watha'iq 'Abdin Files, no. 0069-012050 (March 27, 1947).

27. *Al-Musawwar*, May 30, 1938.

28. 'Umar, *Hadir al-Misriyyin 'Aw Sirr Ta'akhurihum*, 230.

29. 'Umar, *Hadir al-Misriyyin 'Aw Sirr Ta'akhurihum*, 230.

30. More research needs to be done as to why this happened.

31. Ener, *Managing Egypt's Poor*, 107.

32. *al-Ithnayn* (December 10, 1934).

33. Ener, *Managing Egypt's Poor*, 105.

34. *Ministry of the Interior Report for the Year 1922*, TNA, FO 141/586/4.

35. *Nazzaarat al-Dakhiliyya: Qawanin wa-lawa'ih al-Bulis al-Misri* (Cairo: al-Matba'a al-Kubra al-Amiriyya, 1895), 381.

36. *Nazzaarat al-Dakhiliyya*, 539.

37. See Fahmy, *Ordinary Egyptians*, 105–108.

38. *Al-Ahram*, January 11, 1933.

39. "The law on vagrants and on those under surveillance and wanted by the police" (in Arabic), DWQ, *Majlis al-Wizara'* Files, no. 0075-004332 (June 29, 1923).

40. "The law on vagrants and on those under surveillance and wanted by the police."

41. *Al-Ahram*, January 11, 1933.

42. For example, see Robert Tignor, *State, Private Enterprise, and Economic Change in Egypt, 1918–1952* (Princeton: Princeton University Press, 1984), 141–146.

43. For much more on children in Egypt during the first half of the twentieth century, see Heidi Morrison, *Childhood and Colonial Modernity in Egypt* (New York: Palgrave Macmillan, 2015). For a broader examination of destitute children in the Ottoman Empire, see Nazan Maksudyan, *Orphans and Destitute Children in the Late Ottoman Empire* (Syracuse: Syracuse University Press, 2014).

44. Ener, *Managing Egypt's Poor*, 12, 18.

45. Ener, *Managing Egypt's Poor*, 107.

46. Ener, *Managing Egypt's Poor*, 26–30.

47. *Al-Ithnayn*, May 20, 1935.

48. *Al-Ahram*, January 6, 1934.

49. 'Umar, *Hadir al-Misriyyin 'Aw Sirr Ta'akhurihum*, 228–229.

50. Ener, *Managing Egypt's Poor*, 125.

51. Ener, *Managing Egypt's Poor*, 115.

52. *Al-Ahram*, January 14, 1934.

53. *Al-Musawwar*, May 30, 1938.

54. *Al-Musawwar*, May 30, 1938.

55. Henri Lefebvre, *The Production of Space* (Malden: Blackwell, 1991), 63.

56. Lefebvre, *Production of Space*, 4.

Part 2

1. Frederic Courtland Penfield, *Present-Day Egypt* (New York: Century, 1903), 63.

2. Bulis Misr, *Nizam al-Murur bil-Qahirah: Ta'limat li-hafz al-Amn al-'Am* (Cairo: Hikimdar Bulis Misr, 1938), 3.

Chapter 3

1. Richard C. Rath, "Hearing American History," *Journal of American History* 95, no. 2 (2008): 431.

2. Hillel Schwartz, *Making Noise: From Babel to the Big Bang and Beyond* (New York: Zone Books, 2016), 22.

3. Schwartz, *Making Noise*, 430.

4. For the early-nineteenth-century use and importation of horses to Egypt, see Alan Mikhail, *The Animal in Ottoman Egypt* (Oxford: Oxford University Press, 2014), 155–159.

5. André Raymond, *Cairo: City of History* (Cairo: American University in Cairo Press, 2007), 303; John T. Chalcraft, *The Striking Cabbies of Cairo and Other Stories: Crafts and Guilds in Egypt, 1863–1914* (Albany: State University of New York Press, 2005), 59–60.

6. For a nineteenth-century account of the increased use of wheeled vehicles in Alexandria, see Gabriel Charmes, *Five Months at Cairo and in Lower Egypt* (London: Bentley Press, 1883), 21.

7. On Barak, "Scraping the Surface: The Techno-Politics of Modern Streets in Turn-of-Twentieth-Century Alexandria," *Mediterranean Historical Review* 24, no. 2 (2009): 192.

8. For more on Port Said, see Lucia Carminati, "Būr Saʿīd/Port Said, 1859–1914: Migration, Urbanization, and Empire in an Egyptian and Mediterranean Port-City" (PhD diss., University of Arizona, 2018).

9. Khaled Fahmy, "Modernizing Cairo: A Revisionist Narrative," in *Making Cairo Medieval*, ed. Nezar Alsayyad, Irene Bierman, and Nasser Rabbat (Lanham, MD: Lexington Books, 2005), 175.

10. Fahmy, "Modernizing Cairo," 175–185.

11. Fahmy, "Modernizing Cairo," 177.

12. Nancy Reynolds, *A City Consumed: Urban Commerce, the Cairo Fire, and the Politics of Decolonization in Egypt* (Stanford: Stanford University Press, 2012), 29–34.

13. Nezar al-Sayyad, "ʿAli Mubarak's Cairo: Between the Testimony of ʿAlamuddin and the Imaginary of the Khitat," in *Making Cairo Medieval*, ed. Nezar Alsayyad, Irene Bierman, and Nasser Rabbat (Lanham, MD: Lexington Books, 2005), 57.

14. In the small towns and villages of the delta and Upper Egypt, the Provincial Councils had to vote and approve the funds for building agricultural country roads, though they also had to make their case to the Egyptian government for the needed grants and loans in order to modernize their infrastructure. See Ministry of Public Works, Egypt, *Report of the Ministry of Public Works for the Year 1912* (Cairo: Government Press, 1914), 47.

15. Fahmy, "Modernizing Cairo," 178–79.

16. "Cairo Tramway contracts" (in Arabic and French), DWQ, Majlis al-Wizara' Files, no. 0075-059301 (1893–1924).

17. Barak, "Scraping the Surface," 190.

18. William Garstin, *Report upon the Administration of the Public Works Department in Egypt for 1902* (Cairo: National Printing Department, 1903), 50; Ministry of Public Works, *Annual Report for 1921–1922* (Cairo: Government Press, 1925), 185.

19. Aaron George Jakes, "The Scales of Public Utility: Agricultural Roads and State Space in the Era of the British Occupation," in *The Long 1890s in Egypt: Colonial Quiescence, Subterranean Resistance*, ed. Marilyn Booth and Anthony Gorman (Edinburgh: Edinburgh University Press, 2014), 60.

20. Ministry of Public Works, Egypt, *Report of the Ministry of Public Works for the Year 1912*, 39.

21. Joseph Morton Howell, *Egypt's Past, Present and Future* (Dayton, OH: Service, 1929), 308.

22. "City of Alexandria Town Planning Scheme" (October 5, 1923), Records of the Department of State Relating to the Internal Affairs of Egypt, 1910–29 (National Archives Microfilm Publications, Microcopy no. 571, 883:10, Roll 14), National Archives at College Park, College Park, MD.

23. The initial plan was to withdraw British troops from Cairo within four years and from Alexandria within eight years.

24. "Construction of Roads and Railroads Under the Treaty" (23 December 1937), TNA, FO 371/21958.

25. "Construction of Roads and Railroads."

26. "Construction of Roads and Railroads."

27. Stanley Lane-Poole, *The Story of Cairo* (London: J. M. Dent, 1902), 2.

28. For more on Alexandria's tramways, see Barak, "Scraping the Surface," 196.

29. Muhamad Sayyid Kilani, *Tram al-Qahirah: Dirasah Tarikhiyya, Ijtima'iyya, Adabiyya* (Cairo: Matba'it al-Madani, 1968), 16.

30. For more on the importance of trams and trains to the temporality of Egyptian society, see On Barak, *On Time: Technology and Temporality in Modern Egypt* (Berkeley: University of California Press, 2013).

31. For more on the suburbanization of Cairo, see James Moore, "Making Cairo Modern Innovation, Urban Form and the Development of Suburbia, 1880–1922," *Urban History* 41, no. 1 (February 2014): 81–104.

32. For a more detailed list of these early stipulations, see Roland Dussart Desart, "Cairo Tramways: Africa's Greatest Tramway Network, Part 1, *Modern Tramway and Light Rail Transit* 54 (May 1991) 160–161.

33. For some of these classist attitudes in relation to train and tram compartments see *Al-Muqtataf*, September 1908.

34. Ministry of Public Works, *Annual Report for 1899* (Cairo: Government Press, 1900), 270.

35. "Cairo Tramway contracts" (1893–1924).

36. Desart, "Cairo Tramways," Part 1, 164–166.

37. "Cairo Tramway contracts" (1893–1924); Kilani, *Tram al-Qahirah*, 17; Desart, "Cairo Tramways," Part 1, 165.

38. Desart, "Cairo Tramways," Part 1, 165.

39. Moore, "Making Cairo Modern?" 102.

40. Roland Dussart Desart, "The Cairo Tramways: Africa's Greatest Tramway Network," Part 2, *Modern Tramway and Light Rail Transit* 54 (June 1991): 193.

41. The Cairene figures do not include the Heliopolis tram and metro train passengers.

42. Chalcraft, *Striking Cabbies of Cairo*, 59–60.

43. Schwartz, *Making Noise*, 19–23.

44. For an analysis of the acoustical properties of different road surfaces, see C. M. Hogan, "Analysis of Highway Noise," *Water, Air, & Soil Pollution* 2, no. 3 (1973): 387–392.

45. To hear the sound of a mule-drawn cart with creaking wheels in 1928 Cairo, listen to *Cairo street scenes—outtakes*, from 0:28 to 0:36, Fox Movietone News Story 2-18, 1928 [originally filmed on December 28,1928], http://mirc.sc.edu/islandora/object/usc%3A32904.

46. Chalcraft, *Striking Cabbies of Cairo*, 61.

47. See *Cairo street scenes—outtakes*, from 0:24 to 0:33.

48. Lawrence Durrell, *Justine* (New York: Penguin Books, 1985), 236.

49. *Al-Muqatam*, August 1, 1898, cited in Kilani, *Tram al-Qahirah*, 14.

50. For a different analysis of this incident see Barak, *On Time*, 100.

51. This is my translation from Bahija Sidqi Rashid, *Aghani Misr al-Sha'biyya* (Cairo: Matba'it al-Anglu al-Misriyya, 1982), 10–11. Rashid offers an English translation as well, though in order to keep the rhyming quality of the piece, that translation somewhat changes the intended meaning of the song's original lyrics:

> *Al-Kahraba'iyya, al-kahraba'iyya . . . Til 'it tigri 'alah al-Manshiyya*
> *Al-Kahraba'iyya mashiyya kuwayis . . . wa'afuha 'and Khamis*
> *Wa Khamis wa'if mitayyis . . . kab al-bamiyya 'alla al-mulukhiyya*
> *Al-Kahraba'iyya nihas fi nihas . . . wa'afuha 'and 'Ilyyas*
> *Wa 'Ilyyas wa'if mihtas . . . kab al-bamiyya 'ala al-qulqas*

52. Recently, in the early days of hybrid and electric automobiles, the same argument was being made.

53. For more on the sensationalism in the press with regard to this issue, see Barak, *On Time*, 82.

54. Peter Soppelsa, "Urban Railways, Industrial Infrastructure, and the Paris Cityscape, 1870–1914," in *Trains, Culture, and Mobility: Riding the Rails*, ed. Benjamin Fraser and Steven D. Spalding (Lanham, MD: Lexington Books, 2012), 128–129.

55. "Letters from Umar Lutfi Ra'is Majlis Shurah al-Qawanin to the president of Majlis al-Nuzzar" (in Arabic), DWQ, Majlis al-Wizara', no. 0075-015839 (1898).

56. "Letters from Umar Lutfi Ra'is Majlis Shurah al-Qawanin to the president of Majlis al-Nuzzar."

57. See Joel Beinin and Zachary Lockman, *Workers on the Nile: Nationalism, Communism, Islam, and the Egyptian Working Class, 1882–1954* (Cairo: Cairo University Press, 1998).

58. Barak, *On Time*, 56–60.

59. *Al-Dik*, June 17, 1908, quoted in Kilani, *Tram al-Qahirah*, 18.

60. Desart, "Cairo Tramways," Part 1, 165. The Cairo Public Motor Car Service was created by the Cairo Tramway Company, specifically to compete with the SAOC. These early buses were small, ten- to fourteen-passenger "motorized omnibuses."

61. *Al-Dik* (The rooster) was owned and edited by writer and translator Fathi 'Azmi, and according to the Egyptian National Library periodical index (Dar al-Kutub al-Misriyya: Fihris al-Dawriyat al-'Arabiyya Allati Taqtaniha al-Dar), it was in circulation for only one year, from 1907 to 1908.

62. *Annuaire statistique de l'Egypte 1912*, Ministère des finances, Direction de la statistique (Cairo: Imprimerie Nationale, 1912), 474.

63. *Cairo Tanzim Department Annual Report for 1923–24*, Cairo Tanzim Council and Cairo Municipality, TNA, FO 141/633/1.

64. *Cairo Tanzim Department Annual Report for 1923–24*.

65. *Ministry of the Interior Report for the Year 1920*, TNA, FO 141/586/4.

66. E. Homan Mulock, *Report on the Economic and Financial Situation of Egypt*, Department of Overseas Trade (London: His Majesty's Stationary Office, 1923), 32.

67. E. Homan Mulock, *Report on the Economic and Financial Situation of Egypt*, Department of Overseas Trade (London: His Majesty's Stationary Office, 1924), 33.

68. E. Homan Mulock, *Report on the Economic and Financial Situation of Egypt*, Department of Overseas Trade(London: His Majesty's Stationary Office, 1925), 39–40.

69. E. Homan Mulock, *Report on the Economic and Financial Situation of Egypt*, Department of Overseas Trade (London: His Majesty's Stationary Office, 1928), 30.

70. These figures do not include government-owned vehicles. See E. Homan Mulock, *Report on the Economic and Financial Situation of Egypt*, Department of Overseas Trade (London: His Majesty's Stationary Office, 1927), 43; *Annuaire statistique de l'Egypte 1935–36*, Ministère des finances, Direction de la statistique (Cairo: Imprimerie Nationale, 1938); *Annuaire statistique de l'Egypte 1951–54*, Ministère des finances, Direction de la statistique (Cairo: Imprimerie Nationale, 1956), 265–266.

71. Ministry of the Interior, *Annual Report of the Cairo City Police for the Year 1937* (Cairo: Government Press in Bulaq, 1938), 47.

72. Barak, "Scraping the Surface," 198.

73. Ministry of the Interior, *Annual Report of the Cairo City Police for the Year 1937*, 47.

74. Ministry of the Interior, *Annual Report of the Cairo City Police for the Year 1938* (Cairo: Government Press in Bulaq, 1939), 48.

75. Ministry of the Interior, *Annual Report of the Cairo City Police for the Year 1940* (Cairo: Government Press in Bulaq, 1941), 47.

76. Ministry of the Interior, *Annual Report of the Cairo City Police for the Year 1937*, 47.

77. The Frenkel family (of Belarusian Jewish origins) created Frenkel Brothers Productions in Egypt in the mid-1930s. In 1914, the six brothers had moved to Egypt from Jerusalem as children, along with their family. They created the first regular cartoon production in Egypt, using entirely Egyptianized characters and themes. Mishmish Effendi became their most popular cartoon character, after the short cartoon *Mafish Fayda* (There is no use), debuted in 1936 and played in theaters for at least a couple of years. About a dozen Mishmish Effendi cartoons were made. See *Al-Sabah*, May 14, 1937; and Ahmad al-Hadari, *Tarikh al-Sinima fi Misr: al-Guz' al-Thani, 1931–1940* (Cairo: al-Hay'a al-Misriyya al- 'Ammah lil-Kitab, 1989), 305.

78. See *Mafish Mukh*, Frenkel Brothers, 1939, https://www.youtube.com/watch?v=rzUkep3Xifw.

79. Douglas Sladen, *Oriental Cairo: The City of the Arabian Nights* (Philadelphia: Lippincott, 1911), 15.

80. For some of the temporal and cultural implications of railways and tramways in Egypt, see Barak, *On Time*.

81. The original railway station was built in 1856 and entirely rebuilt in 1892. The station building was upgraded in 1955 and 2011. The square where the station is located was first called Bab al-Hadid Square (metal gate square), named after one of the original medieval Cairo gates. After 1955, when Nasser erected an original statue of Ramses II in the square, most Egyptians referred to it as Ramses Square.

82. For an excellent history of cabbies in urban Egypt, see Chalcraft, *Striking Cabbies of Cairo*.

83. For much more on the role of urban commerce in Ataba Square, see Reynolds, *A City Consumed*, 29–34.

84. See, for example, the 1959 comedy film *al-Ataba al-Khadra*, directed by Fatin Abd al-Wahab and starring Ismail Yasin, Ahmad Mazhar, and Sabah. In the movie, Mabruk (played by Yasin) is an Upper Egyptian peasant who arrives in Cairo after selling all his land and is initially defrauded through being persuaded to buy al-Ataba al-Khadra Square for 12,000 LE. In the early organized comedic theater in Egypt, the character of Kishkish Bey, performed by the comedic icon Naguib al-Rayhani in dozens of plays, was a rich countryside peasant who sold his land and arrived to the city, only to be tempted by its many distractions.

85. I thank Walter Armbrust for pointing out to me the importance of anonymity as experienced by newcomers to the city.

86. For an excellent overview and critical examination of *Bab al-Hadid*, see Joel Gordon, "Broken Heart of the City: Youssef Chahine's *Bab al Hadid (Cairo Station)*," *Journal for Cultural Research* 16, no. 2–3 (2012): 217–237.

87. See *Bab al-Hadid* [Cairo station], directed by Youssef Chahine (Cairo, 1958). The movie's main protagonist, Qinawi (played by Chahine) is a tragic figure who has recently migrated from Upper Egypt to seek his fortune in the big city. At the beginning of the movie, Qinawai is found crouched over, destitute and hungry, by the station's principle newspaper agent (who is also the film's narrator). Feeling sympathy for Qinawi, the agent hires him as a paper seller. Despite his employment, Qinawi is unable to cope psychologically with life in Cairo. His sexual frustration and cultural alienation lead to his institutionalization, for insanity and attempted murder. Although sensationalized and exaggerated, the movie's plot reflected real anxieties about many of the changes taking place in Egypt at the middle of the twentieth century.

88. David Nye, *Electrifying America: Social Meanings of a New Technology* (Cambridge, MA: MIT Press, 1992), ix–x.

89. For an excellent examination of the affordability and availability of trams, see On Barak, "Scraping the Surface," 195–197.

90. *Cairo Tanzim Department Annual Report for 1923–24*. Because electricity and coal prices were high during World War I, the government then allowed tramway companies to temporarily raise their prices to 6 and 12 millimes. When electricity prices decreased in 1923, negotiations between the Tramway Company and the government led

to a compromise of keeping the prices at a higher rate "on condition that a percentage of the gross receipts should be paid annually to [the] Government." These extra funds were critical in paying for road infrastructure throughout the 1920s and 1930s. For example, they paid for building al-Azhar Street and Prince Faruq Road, which were finished in 1926 (see Chapters 1 and 2).

91. This street-level reality, continued until the 1930s and 1940s, when women's public role expanded widely among all classes.

92. Despite the overwhelming presence of all these women in the streets and public squares, intellectual discourses and press editorials during the first two decades of the twentieth century primarily discussed the role of elite women and their veiled or unveiled (potential) entry into the public sphere. Even today, the common belief in Egypt is that the feminists Hoda Sharawi and Sayza Nabarawi were the first Egyptian women to remove their face veils, upon their return from the 1923 International Woman Suffrage Alliance conference in Rome—marking the official entry of Egyptian women into the public sphere. It is as if the hundreds of thousands of Egyptian working-class women and peasants, some of whom never wore a face veil to begin with, simply did not exist. This clear gap between early-twentieth-century reality and the idealized and imagined gender and class segregationist vision of the elites was most visible in Egypt's city streets. For an excellent examination of working-class Egyptian women during the first half of the twentieth century, see Hanan Hammad, *Industrial Sexuality: Gender, Urbanization, and Social Transformation in Egypt* (Austin: University of Texas Press, 2016).

93. James Ryan, "'Unveiling' the Tramway: The Intimate Public Sphere in Late Ottoman and Republican Istanbul," *Journal of Urban History* 44, no. 5 (2016): 1–24; Lauren Berlant, *The Female Complaint: The Unfinished Business of Sentimentality in American Culture* (Durham, NC: Duke University Press, 2008). See also Alev Cinar, *Modernity, Islam, and Secularism in Turkey: Bodies, Places, and Time* (Minneapolis: University of Minnesota Press, 2005).

94. Barak, *On Time*, 160–175.

95. With the exception of the Giza line heading westward toward the pyramids and the Heliopolis line, most Cairene tramways kept relatively low speeds.

96. According to On Barak, at the turn of the twentieth century, in order to save time (and money), tram "drivers tended to stop or deaccelerate near stations only for European passengers or for women, while local Egyptian men were forced to jump on and off moving trams." Barak, *On Time*, 10.

97. Hector Dinning, *By-ways on Service: Notes from an Australian Journal* (London: Constable, 1918), 40.

98. Dinning, *By-ways on Service*, 40.

99. *Ministry of the Interior Report for the Year 1922*.

100. Richard Carrington, *The Tears of Isis: The Story of a New Journey from the Mouth to the Source of the Nile* (London: Chatto & Windus, 1959), 76.

101. *Ministry of the Interior Report for the Year 1920*.

102. *Ministry of the Interior Report for the Year 1920*.

Chapter 4

1. Fathi Radwan, *Khati al-Ataba: Hayyat Tifl Misri* (Cairo: Dar al-Ma'arif, 1973), 95.

2. For the example of the politics and business of electrification in Palestine under the mandate, see Ronen Shamir, *Current Flow: The Electrification of Palestine* (Stanford: Stanford University Press, 2013); and Fredrik Meiton, *Electrical Palestine: Capital and Technology from Empire to Nation* (Berkeley: University of California Press, 2019).

3. "Iltimas min ashab mahalat al-hilaqah bil-tasrih lahum bi'adam itfa' al-nur qabl al-sa'a al-tasi'a masa'an" [An appeal from barbershop owners to allow them to keep the lights on until 9:00 p.m.], DWQ, Majlis al-Wizara' Files, no. 0075-30000 (August 21, 1917).

4. "Iltimas min ashab mahalat al-hilaqah bikum al-Shakafah" [An appeal from barbershop owners in Kum al-Shakafah], DWQ, Majlis al-Wizara' Files, no. 0075-003369 (August 18, 1917).

5. For more on rationing, inflation, and high prices during World War I in Egypt, see Ziad Fahmy, *Ordinary Egyptians: Creating the Nation through Popular Culture* (Stanford: Stanford University Press, 2011), 115–120; for more on Egypt during World War I, see Kyle Anderson, "The Egyptian Labor Corps: Logistical Laborers in War and Revolution, 1914–1919" (PhD diss., Cornell University, 2017).

6. "Iltimas min ashab mahalat al-hilaqah bil-tasrih lahum bi'adam itfa' al-nur qabl al-sa'a al-tasi'a masa'an" [An appeal from barbershop owners to allow them to keep the lights on until 9:00 p.m.].

7. Martin S. Briggs, *Through Egypt in Wartime* (London: Frederick Stokes, 1918), 13, 25–26.

8. For much more on Cairene department stores and consumerism, see Nancy Reynolds, *A City Consumed: Urban Commerce, the Cairo Fire, and the Politics of Decolonization in Egypt* (Stanford: Stanford University Press, 2012).

9. Wolfgang Schivelbusch, *Disenchanted Night: The Industrialization of Light in the Nineteenth Century* (Berkeley: University of California Press, 1995), 61.

10. Schivelbusch, *Disenchanted Night*, 63–64.

11. "Nizarat al-Ashghal al-'Umumiyya: Awraq 'An Tawlid al-Kahraba wa Bina' Shawari' al-Qahirah" [Ministry of Public Works: Papers concerning road building and electricity generation in Cairo], DWQ, Majlis al-Wizara' Files, no. 0075-059302 (1905–1915). Contracts between the Egyptian government and the Lebon Company are included in this file.

12. "Nizarat al-Ashghal al-'Umumiyya: Awraq 'An Tawlid al-Kahraba wa Bina' Shawari' al-Qahirah" [Ministry of Public Works: Papers concerning road building and electricity generation in Cairo].

13. See, for example, "Nizarat al-Ashghal al-'Umumiyya: Mashru' li-Insha' Ma'mal lit-awlid al-Kahrabah li-maslahit al-sikah al-hadid fi al-Qahirah" [Ministry of Public Works: The building of an electric plant for the Railways Department in Cairo], DWQ, Majlis al-Wizara' Files, no. 0075-049490 (May 15, 1923).

14. "Nizarat al-Ashghal al-'Umumiyya: Awraq 'An Tawlid al-Kahraba wa Bina' Shawari' al-Qahirah" [Ministry of Public Works: Papers concerning road building and

electricity generation in Cairo]. Article 10 also agrees to allow the tram companies to supply the electrical power necessary to open and close the moveable automatic bridges [*ponts tournants*].

15. Robert Vitalis, *When Capitalists Collide: Business Conflict and the End of Empire in Egypt* (Berkeley: University of California Press, 1995), 148–149.

16. Ministry of Public Works, *Annual Report for 1921–1922* (Cairo: Government Press, 1925), 14.

17. In its 1932 negotiations with the Lebon Company, the Egyptian government insisted on reducing the cost of electricity to 17 millimes per kilowatt-hour, but a satisfactory agreement could not be reached.

18. A critical part of the formula for the gas and electricity prices charged by Lebon in the Egyptian market was the pegging of the price to the Parisian market, but apparently Lebon was playing with the numbers to charge higher prices in Egypt. "Lebon & Co. and Alexandria" (1934), TNA, FO 141/482/3.

19. Vitalis, *When Capitalists Collide*, 148–149.

20. A. N. Cumberbatch, *Egypt: Economic and Commercial Conditions in Egypt*, Overseas Economic Surveys series (London: His Majesty's Stationery Office, 1952), 13.

21. Cumberbatch, *Egypt: Economic and Commercial Conditions in Egypt* (1952), 13–14.

22. Vitalis, *When Capitalists Collide*, 148–149.

23. Galal Amin, *Madha 'Alamatni al-Haiyyah: Sirah Dhatiyya* (Cairo: Dar al-Shuruq, 2007), 45.

24. Ministry of Public Works, *Annual Report for 1921–1922*, 219.

25. G. H. Selous, *Report on Economic and Commercial Conditions in Egypt*, Department of Overseas Trade (London: His Majesty's Stationery Office, 1935), 34; C. Empson, *Report on Economic and Commercial Conditions in Egypt*, Department of Overseas Trade (London: His Majesty's Stationery Office, 1939), 25.

26. J. W. Taylor, *Egypt: Economic and Commercial Conditions in Egypt*, Overseas Economic Surveys series (London: His Majesty's Stationery Office, 1948), 42; for more on Egyptian manufacturing and the business climate in interwar Egypt, see Eric Davis, *Challenging Colonialism: Bank Misr and Egyptian Industrialization, 1920–1941* (Princeton: Princeton University Press, 1983).

27. Empson, *Report on Economic and Commercial Conditions in Egypt* (1939), 26.

28. Selous, *Report on Economic and Commercial Conditions in Egypt* (1935), 38; Cumberbatch, *Egypt: Economic and Commercial Conditions* (1952), 40.

29. By the 1970s, portable cassette tape players, many of which were equipped with radio receivers, replaced radios as the ultimate symbol of middle-class upward mobility. See Andrew Simon, "Sounding History: Cassettes, Culture, and Everyday Life in Modern Egypt" (PhD diss., Cornell University, 2017).

30. Cumberbatch, *Egypt: Economic and Commercial Conditions* (1952), 66.

31. UNESCO, *Statistics on Radio and Television 1950–1960* (Paris: United Nations Educational, Scientific and Cultural Organization, 1963), 61. These figures are likely understated as they reflect only officially licensed radio sets.

32. Hector Dinning, *By-ways on Service: Notes from an Australian Journal* (London: Constable, 1918), 38.

33. Khaled Fahmy, "Modernizing Cairo: A Revisionist Narrative," in *Making Cairo Medieval*, ed. Nezar Alsayyad, Irene Bierman, and Nasser Rabbat (Lanham, MD: Lexington Books, 2005), 183. There were exceptions to these laws during pilgrimages, *mawalid*, and major religious holidays, in order to accommodate the increase in the number of people visiting or passing through Cairo.

34. *Nazzaarat al-Dakhiliyya: Qawanin wa-lawa'ih Al-Bulis al-Misri* (Cairo: al-Matba'a al-Kubra al-Amiriyya, 1895), 370–373.

35. *Tarjamat La'iha: Tata'alaq bi Isti'mal al-Afrad al-Turuq al-'Umumiyya* (Cairo: al-Matba'a al-Kubrah al-'Amiriyya, 1903); *Nazzaarat al-Dakhiliyya*, 375.

36. For more on the use and abuse of the capitulations that allowed foreign nationals to be tried in consular courts rather than Egyptian courts, see Will Hanley, *Identifying with Nationality: Europeans, Ottomans, and Egyptians in Alexandria* (New York: Columbia University Press, 2017); Ziad Fahmy, "Jurisdictional Borderlands: Extraterritoriality and 'Legal Chameleons' in Precolonial Alexandria, 1840–1870," *Comparative Studies in Society and History* 55, no. 2 (April 2013): 305–329.

37. *Buza* is considered a "low-class" Egyptian alcoholic drink and is typically made from fermented bread. Ali Pasha Mubarak, *Al-Khitat al-Tawfiqiyya al-Jadida li-Misr al-Qahira wa-Mudunaha wa-Biladaha al-Qadima wa al-Shahira* (Cairo: Matba'at Dar al-Kutub wa al-Watha'iq al-Qawmiyya bil-Qahira, 2004 [1887/88]), 1: 218, 238. For more on Egyptian *buza* bars, see Fahmy, *Ordinary Egyptians*, 159.

38. For an examination of Middle Eastern coffee shop culture, see Kirli Cengiz, "The Struggle over Space: Coffeehouses of Ottoman Istanbul, 1780–1845" (PhD diss., State University of New York at Binghamton, 2000); and Omar Carlier, "Le café maure: Sociabilité masculine et effervescence citoyenne (Algérie XVIIe–XXe siècles)," *Annales: Histoire, Sciences Sociales* 45, no. 4 (1990): 975–1004.

39. The name *guzah* is derived from the Arabic *guz al-hind* (coconut). *Guzahs* are small handheld pipes, traditionally made of a coconut shell, where the tobacco is held and smoked through an attached hollow wooden pipe. The more common variety that was available at the turn of the century was made of a tin base and had a wooden pipe arm.

40. Radwan, *Khati al-Ataba*, 60–61.

41. Radwan, *Khati al-Ataba*, 60–61.

42. *al-Muftah*, January 15, 1906.

43. For much more detail on Cairene coffeehouses, see Alon Tam, "Cairo's Coffeehouses in the Late Nineteenth- and Early Twentieth-Centuries: An Urban and Socio-Political History" (PhD diss., University of Pennsylvania, 2018).

44. Fahmy, *Ordinary Egyptians*, 132; *Al-Masrah*, July 25, 1927; Naguib al-Rihani, *Mudhakirat Naguib al-Rihani* (Cairo: Dar al-Hilal, 1959), 26–27; Ibrahim Ramzi, *Masrahuna Ayyam Zaman wa Tarikh al-Fananin al-Qudama* (Cairo: Matba'at al-Salam, 1984), 25; Badi' Khayri, *Mudhakirat Badi' Khayri: Khamsah wa Arbi'un Sana 'ala Adwa' al-Masrah* (Beirut: Dar al-Thakafa, n.d.), 10.

45. For more on Egyptian café culture as it related to orality and aurality, reading aloud, and literature and politics, see Fahmy, *Ordinary Egyptians*, 35–36, 145–147, 155–156; and "Intelligence Report on the Egyptian Situation" (May 5, 1919), TNA, FO 141/781/8915. This report documents how illicit circulars during the 1919 Revolution were spread through urban coffee shops.

46. Fahmy, *Ordinary Egyptians*, 35–36; H. Hamilton Fyfe, *The New Spirit in Egypt* (London: William Blackwood, 1911), 113. In French, backgammon is called *trictrac*, an obvious onomatopoeia. For the sounds of a backgammon game in Cairo listen to https://www.youtube.com/watch?v=DQ8QDgU3Sik&list=PLQx9s9aIVIWNFMuUMkbUoHHiGYMtmG9gk&index=3.

47. For nineteenth-century examples of some of these musical and theatrical performances, see Adam Mestyan, *Arab Patriotism: The Ideology and Culture of Power in Late Ottoman Egypt* (Princeton: Princeton University Press, 2017).

48. For an excellent summary of Cairene nightlife during the first third of the twentieth century, see Heather D. Ward, *Egyptian Belly Dance in Transition: The Raqs Sharqi Revolution, 1890–1930* (Jefferson: McFarland, 2018), 23–46.

49. For more on these private stations, see Ziad Fahmy, "Early Egyptian Radio: From Media-Capitalism to Media-Etatism," in "Media Transitions and Cultural Production in Arab Societies: Trans-historical Perspectives," ed. Barbara Winckler and Teresa Pepe, special issue, *Middle East Journal of Culture and Communication*, forthcoming.

50. See, for example, a frequently printed *al-Ahram* radio advertisement titled "Why Hesitate?" The ad, sponsored by Star Orient Stores, specifically highlights the store's installment plans. *Al-Ahram*, July 2, 1933 and July 4, 1933.

51. *Al-Ahram*, July 12, 1933.

52. *Al-Ahram*, July 5, 1933.

53. *Al-Ahram*, July 12, 1933.

54. *Al-Ahram*, July 12, 1933.

55. *Al-Ahram*, July 12, 1933.

56. *Al-Ahram*, July 14, 1933.

57. *Al-Ahram*, July 14, 1933.

58. Naguib Mahfouz, *Midaq Alley* (New York: Anchor Books, 1992 [1947]), 3. The first few pages of the novel also directly address the de facto obsolescence of storytellers, as the owner of the café bans the older storyteller from sitting in his café.

59. For more on amplification technology and the early sounded motion pictures, see Emily Thompson, "Wiring the World: Acoustical Engineers and the Empire of Sound in the Motion Picture Industry, 1927–1930," in *Hearing Cultures: Essays on Sound, Listening and Modernity*, ed. Veit Erlmann (New York: Berg, 2004), 192–197.

60. Douglas Sladen, *Oriental Cairo: The City of the Arabian Nights* (Philadelphia: Lippincott, 1911), 109.

61. See Francesca Biancani, "International Migration and Sex Work in Early Twentieth-Century Cairo," in *A Global Middle East: Mobility, Materiality and Culture in the*

Modern Age, 1880–1940, ed. Liat Kozma, Cyrus Schayech, and Avner Wishnitzer (New York: I. B. Tauris, 2014), 117–118. Wajh al-Birkah Street was usually called Wish al-Birkah Street in vernacular Egyptian.

62. Imad al-Din Street runs on a north–south axis two blocks west of Azbakiyya Garden, just west of the Continental and Shepherd Hotels.

63. For more on brothels and criminality in Alexandria, see Nefertiti Takla, "Murder in Alexandria: The Gender, Sexual and Class Politics of Criminality in Egypt, 1914–1921" (PhD diss., University of California, Los Angeles, 2016).

64. Briggs, *Through Egypt in War-Time*, 13.

65. Eustace Reynolds-Ball, *Cairo of Today: A Practical Guide to Cairo and the Nile* (London: A & C Black, 1916), 12–16. For a vivid and loud description of the Café Egyptien during World War I, see A. W. Keown, *Forward with the Fifth: The Story of Five Years' War Service, Fifth Inf. Battalion, A.I.F.* (Melbourne: Specialty Press, 1921), 64–65.

66. Fahmy, *Ordinary Egyptians*, 122–130. For more on the theater of Ali al-Kassar, see Eve M. Troutt Powell, *A Different Shade of Colonialism: Egypt, Great Britain, and the Mastery of the Sudan* (Berkeley: University of California Press, 2003).

67. Lothaire Loewenbach, *Promenade Autour de l'Afrique, 1907* (Paris: Flammarion, 1908), 219–220.

68. Karin van Nieuwkerk, *A Trade like Any Other: Female Singers and Dancers in Egypt* (Austin: University of Texas Press, 1995), 43–45, 49.

69. Loewenbach, *Promenade Autour de l'Afrique*, 219–220.

70. Loewenbach, *Promenade Autour de l'Afrique*, 219–220.

71. I was able to locate four different photo postcards of the Eldorado (ca. 1905–1908). All four feature bottles of alcohol on the stage.

72. Hector Dinning, *Nile to Aleppo, with the Light-Horse in the Middle-East* (New York: Macmillan, 1920), 237–238.

73. Dinning, *Nile to Aleppo*, 239–241.

74. Radwan, *Khati al-Ataba*, 132; Dinning, *Nile to Aleppo*, 239–240.

75. Radwan, *Khati al-Ataba*, 132.

76. For more on these traveling theater troupes, and particularly Rihani and Kassar impersonators, see Fahmy, *Ordinary Egyptians*, 128.

77. Dinning, *Nile to Aleppo*, 243.

78. Schivelbusch, *Disenchanted Night*, 142.

79. Schivelbusch, *Disenchanted Night*, 60.

80. Schivelbusch, *Disenchanted Night*, 55.

81. Ministry of Public Works, Egypt, *Report of the Ministry of Public Works for the Year 1912* (Cairo: Government Press, 1914), 29.

82. David Nye, *Electrifying America: Social Meanings of a New Technology* (Cambridge, MA: MIT Press, 1992), ix.

83. For more on Heliopolis and Luna Park, see Robert Ilber, *Heliopolis: Le Caire, 1905–1922* (Paris: Centre National de la Recherche Scientifique, 1981).

84. Dinning, *Nile to Aleppo*.

85. E. Homan Mulock, *Report on the Economic and Financial Situation of Egypt*, Department of Overseas Trade (London: His Majesty's Stationary Office, 1924), 29.

86. Cumberbatch, *Egypt: Economic and Commercial Conditions in Egypt* (1952), 66.

87. For more on Masabni, see Roberta L. Dougherty, "Badi'a Masabni, Artiste and Modernist: The Egyptian Print Media's Carnival of National Identity," in *Mass Mediations: New Approaches to Popular Culture in the Middle East and Beyond*, ed. Walter Armbrust (Berkeley: University of California Press, 2000), 243–268; Virginia Danielson, "Artists and Entrepreneurs: Female Singers in Cairo during the 1920s," in *Women in Middle Eastern History: Shifting Boundaries in Sex and Gender*, ed. Nikki Keddie and Beth Baron (New Haven: Yale University Press, 1991), 292–309.

88. van Nieuwkerk, *A Trade like Any Other*, 42; Ward, *Egyptian Belly Dance in Transition*, 24, 55.

89. Hanafi Mahallawi, *Layali al-Qahirah fi 'Asr Badi'a: al-Raqisah Allati Hakamat Bashawat Misr wa-Sa'aliq 'Imad al-Din* (Cairo: Dar al-Kitab al- 'Arabi, 1997), 111.

90. Badi'a Masabni and Nazik Basilah, *Mudhakarat Badi'a Masabni* (Beirut: Dar Maktabit al-Hayah 1960), 336–340.

91. Thompson, "Wiring the World," 193–194.

92. *Al-Ahram*, November 9, 1896; Ali Abu Shadi, *Waqa'i' al-Sinima al-Misriyya* (Cairo: al-Hayyi'a al-Misriyya al- 'Amma lil-Kitab, 2003), 13–17.

93. Shadi, *Waqa'i' al-Sinima al-Misriyya*, 23.

94. For a more detailed look at early Egyptian cinemas, see Viola Shafik, *Arab Cinema: History and Cultural Identity* (Cairo: American University in Cairo Press, 1998), 10–15.

95. Alfred Cunningham, *Today in Egypt: Its Administration, People and Politics* (London: Hurst, 1912), 35.

96. Briggs, *Through Egypt in War-Time*, 34–35.

97. "Letter from Geist to Howell" (March 18, 1926), Records of the Department of State Relating to the Internal Affairs of Egypt, 1910–29 (*National Archives Microfilm Publications*, Microcopy No. 571, 883:40, Roll 15), National Archives at College Park, College Park, MD.

98. The Josy Film Agency report was quoted in a letter dated April 8, 1926, and written by North Winship, the American Consul General in Alexandria. See Records of the Department of State Relating to the Internal Affairs of Egypt, 1910–29 (*National Archives Microfilm Publications*, Microcopy No. 571, 883:40, Roll 15), National Archives at College Park, College Park, MD.

99. Radwan, *Khati al-Ataba*, 130.

100. For more on this technological transition and its impact on the soundscapes of modernity, see Emily Thompson, *The Soundscape of Modernity: Architectural Acoustics and the Culture of Listening in America, 1900–1933* (Boston: MIT Press, 2004).

101. Thompson, "Wiring the World," 193–195.

102. *Epigram Magazine*, August 15, 1929, quoted in Thompson, "Wiring the World," 195.

103. G. H. Selous and L.B.S. Larkins, *Report on Economic and Commercial Conditions in Egypt*, Department of Overseas Trade (London: His Majesty's Stationery Office, 1933), 83.

104. Selous, *Report on Economic and Commercial Conditions in Egypt* (1935), 49.

105. Empson, *Report on Economic and Commercial Conditions in Egypt* (1939), 19.

106. The first Egyptian movie theater to be air-conditioned was Metro Cinema, in February 1940. The theater could seat over 1,500 people. See Ahmad al-Hadari, *Tarikh al-Sinima fi Misr: al-Guz' al-Thani, 1931–1940* (Cairo: al-Hay'a al-Misriyya al-'Ammah lil-Kitab, 1989), 416.

107. Taylor, *Egypt: Economic and Commercial Conditions in Egypt* (1948), 18.

108. *Al-Sabah*, August 1, 1946.

109. Selous and Larkins, *Report on Economic and Commercial Conditions in Egypt* (1933), 84.

110. Selous and Larkins, *Report on Economic and Commercial Conditions in Egypt* (1933), 83.

111. Shafik, *Arab Cinema*, 12–13.

112. G. H. Selous, *Report on Economic and Commercial Conditions in Egypt*, Department of Overseas Trade (London: His Majesty's Stationery Office, 1937), 54.

113. Taylor, *Egypt: Economic and Commercial Conditions in Egypt* (1948), 86.

114. Thompson, "Wiring the World," 207.

115. For an excellent recent examination of Egypt's transition to talkies, see Ifdal Elsaket, "Sound and Desire: Race, Gender, and Insult in Egypt's First Talkie," *International Journal of Middle East Studies* 51, no. 2 (2019): 203–232.

116. Schivelbusch, *Disenchanted Night*, 73.

117. For a cultural history of the microphone in Egypt in relation to the singer 'Abd al-Halim Hafiz, see Martin Stokes, "'Abd al-Halim's Microphone," in *Music and the Play of Power in the Middle East, North Africa and Central Asia*, ed. Laudan Nooshin (New York: Routledge, 2016), 55–74.

Part 3

1. *Nazzaarat al-Dakhiliyya: Qawanin wa-lawa'ih al-Bulis al-Misri* (Cairo: al-Matba'a al-Kubra al-Amiriyya, 1895), 379.

2. *The Egyptian Gazette*, March 17, 1924. The *Gazette*'s report covers the opening of parliament two days earlier, on March 15, 1924.

Chapter 5

1. Loudspeakers were in use in Egypt from the mid-1930s on, but at first, they were used mostly by the state (see Chapter 6).

2. For an in-depth analysis of the cultural and legal evolution of Egyptian marriages during the nineteenth and twentieth centuries, see Kenneth Cuno, *Modernizing Marriage: Family, Ideology, and Law in Nineteenth- and Early Twentieth-Century Egypt* (Syracuse: Syracuse University Press, 2015); and Hanan Kholoussy, *For Better, for Worse: The Marriage Crisis That Made Modern Egypt* (Stanford: Stanford University Press, 2010).

3. For an excellent analysis of these tensions as manifested in anti-*mawalid* sentiments, see Samuli Schielke, "Hegemonic Encounters: Criticism of Saints-Day Festivals and the Formation of Modern Islam in Late 19th and Early 20th-Century Egypt," in *Die Welt des Islams* 47, no. 3/4 (2007): 319–355; and Meir Hatina, "Religious Culture Contested: The Sufi Ritual of Dawsa in Nineteenth-Century Cairo," *Die Welt des Islams* 47, no. 1 (2007): 33–62.

4. Bruce Smith, "Listening to the Wild Blue Yonder: The Challenge of Acoustic Ecology," in *Hearing Cultures: Essays on Sound, Listening and Modernity*, ed. Veit Erlmann (New York: Berg, 2004), 35–36.

5. To listen to an Egyptian style *zaghruta*, play https://soundcloud.com/rsvp/zaghareet-sample-from-egypt.

6. Evelyn Early, *Baladi Women of Cairo: Playing with an Egg and a Stone* (London: Lynne Reimer, 1993), 105. For a view from another middle-eastern society, see Jennifer E. Jacobs, "Ululation in Levantine Society: The Cultural Reproduction of an Affective Vocalization" (PhD diss., University of Pennsylvania, 2008).

7. Frederic Courtland Penfield, *Present-Day Egypt* (New York: Century, 1903), 21.

8. Early, *Baladi Women of Cairo*, 5.

9. Constance Fenimore Woolson, *Mentone, Cairo and Corfu* (New York: Harper & Brothers, 1896), 262.

10. For a description of a 1910 Cairene *shadir*, richly decorated and equipped with electric light, see Douglas Sladen, *Oriental Cairo: The City of the Arabian Nights* (Philadelphia: Lippincott, 1911), 259.

11. *Tarjamat La'iha: Tata'alaq bi Isti'mal al-Afrad al-Turuq al-'Umumiyyah* (Cairo: al-Matba'a al-Kubrah al-'Amiriyya, 1903); *Nazzaarat al-Dakhiliyya: Qawanin wa-lawa'ih al-Bulis al-Misri* (Cairo: al-Matba'a al-Kubra al-Amiriyya, 1895), 373.

12. *Tarjamat La'iha*.

13. Fathi Radwan, *Khati al-Ataba: Hayyat Tifl Misri* (Cairo: Dar al-Ma'arif, 1973), 87.

14. For more on the creation of a modern Egyptian army under Muhammad Ali, see Khaled Fahmy, *All the Pasha's Men: Mehmed Ali, His Army and the Making of Modern Egypt* (Cambridge: Cambridge University Press, 1997).

15. For a brief history of military bands under Muhammad Ali in Egypt, see Adam Mestyan, "Sound, Military Music, and Opera in Egypt during the Rule of Mehmet Ali Pasha (r. 1805–1848)," in *Ottoman Empire and European Theatre*, vol. 2 of *The Time of Joseph Haydn—From Sultan Mahmud I to Mahmud II (r. 1730–1839)*, ed. Michael Huttler and Hans Ernst Weidinger (Vienna: Hollitzer Wissenschaftsverlag, 2014), 631–656.

16. Mestyan, "Sound, Military Music, and Opera in Egypt," 649.

17. Karin van Nieuwkerk, *A Trade like Any Other: Female Singers and Dancers in Egypt* (Austin: University of Texas Press, 1995), 50.

18. Salwa El-Shawan, "Western Music and Its Practitioners in Egypt (ca. 1825–1985): The Integration of a New Musical Tradition in a Changing Environment," *Asian Music* 17, no. 1 (1985): 143–153.

19. For more detail on the early history of the Egyptian army, see Fahmy, *All the Pasha's Men*.

20. See, for example, Gabriel Charmes, *Five Months at Cairo and in Lower Egypt* (London: Bentley Press, 1883), 101.

21. Alfred Cunningham, *Today in Egypt: Its Administration, People and Politics* (London: Hurst, 1912), 34.

22. See the television interview of 'Izat al-Faiyumi, one of the last surviving members of the original Hasaballah Band. He joined the band as a seventeen-year-old in 1935. According to Faiyumi, at that time the Hasaballah Band was run by haj Ali Hasaballah, the son of the original founder. In 1970, al-Faiyumi started his own Hasaballah Band, which he named himself. The band was still operating in 2017, when the interview was filmed. *Firqit Hasaballah*, Al-Jazeera documentary (2017), https://www.youtube.com/watch?v=ZqkYTc_wYdQ.

23. For an excellent history of Muhammad Ali Street and its role in twentieth-century Egyptian celebrations and weddings, see van Nieuwkerk, *A Trade like Any Other*; also see Katherine Zirbel, "Musical Discursions: Spectacle, Experience and Political Economy among Egyptian Performers in Globalizing Markets" (PhD diss., University of Michigan, 1999).

24. van Nieuwkerk, *A Trade like Any Other*, 50.

25. See *Firqit Hasaballah*.

26. Muhammad Ghunaym, *Al-'Adat wa al-Taqalid fi Dilta Misr* (Cairo: 'Ayn lil-Dirasat wa al-Buhuth al-Insaniyya wa al-Ijtima'iyya, 2005), 474, 491.

> *Rushuh al-shari' maya . . . Da 'arusit al-ghali gayah*
> *Rushuh al-shari' kukakulah [or kazuzah] . . . Da 'arusit al-ghali manqulah*
> *Rushuh al-shari' riha . . . Da 'arusit al-ghali maliha*
> *Rushuh al-shari' kahrabah . . . Ya 'arusit al-ghali marhabah*
> *Aywah ya harah dalmah . . . Ihna 'illi nawrnaha*
> *Aywah ya harah huss-huss . . . Ihna 'illi 'amalnalkum hiss*

27. van Nieuwkerk, *A Trade like Any Other*, 24–25.

28. Street *shawadir* supplied with electric lights existed in Cairo as early as 1910. See, for example Sladen, *Oriental Cairo*, 259.

29. Karl Baedeker, *Egypt and the Sudan: Handbook for Travelers* (Leipzig: 1914), xciii, 62.

30. Brides were also transported in elaborate litters on camelback, until carriages, horse-drawn carts, and later on automobiles gradually replaced them. For an 1899 account of the use of camel-transported kettledrums, see *Century Magazine*, October 1899.

31. For an excellent 1980 film depiction of a circumcision *zaffah*, in the village of al-Haraniyya outside Cairo, see *Faqrah min Shari' Muhammad 'Ali: Zaffit Mutahir fi al-Haraniyya* (Burnamij al-Shari' al-Misri [Egyptian Television], 1980), https://www.youtube.com/watch?v=n5t3SJMWcI4.

32. Abd al-Hakim Khalil al-Sayid Ahmad, '*Adat wa Taqalid al-Zawaj: Dirasah fi al-Thaqafah al-Sha'biyyah* (Cairo: Misr al-'Arabiyya lil-Dirasah wa al-Tawdhi, 2014), 125.

33. Fathi al-Sanfawi, *Aghani al-Hub wa al-Zawaj wa al-Afrah* (Cairo: Starburs lil-Tiba'a wa al-Nashr, 2009), 96–97.

34. For a description of an early-twentieth-century pilgrimage and a groom *zaffah*, see Sladen, *Oriental Cairo*, 262; al-Sanfawi, *Aghani al-Hub wa al-Zawaj wa al-Afrah*, 96–97.

35. Sladen, *Oriental Cairo*, 262.

36. al-Sanfawi, *Aghani al-Hub wa al-Zawaj wa al-Afrah*, 105–106; van Nieuwkerk, *A Trade like Any Other*, 24, 50, 56.

37. "*Fayn gihazik ya arussah faraguni 'alayh . . . Fin magayib abuki Allah yinawar 'alayh.*" Al-Sanfawi, *Aghani al-Hub wa al-Zawaj wa al-Afrah*, 185.

38. *Cook's Handbook for Egypt and the Egyptian Sudan* (London: Thomas Cook, 1921), 111.

39. Muhammad Umran, *Al-Ghina' al-Fulkluri* (Cairo: al-Markaz al-Misri Lil-Thakafah wa al-Funun, 2012), 53.

40. al-Sanfawi, *Aghani al-Hub wa al-Zawaj wa al-Afrah*, 163.

'Ala'it 'amis-ha . . . labisit 'amis-ha . . . rahit li'aris-ha . . . bilayl al-sa'a wahda
Tali' yinaghis-ha nazil yinaghis-ha . . . kassar ghawish-ha . . . bilayl al-sa'a wahda
Tali' wayaha . . . nazil wayaha . . . wa'i' fi hawaha . . . bilayl al-sa'a wahda

41. Muhammad Shabana, "Aghani al-Zawaj fi Ba'd Qurah al-Mahalah al-Kubrah," in *Al-A'ras al-Sha'bia* (Cairo: Dar al-Za'im lil-Tiba'ah al-Hadithah, 2005), 155.

'al sirir al-hayla hub . . . ya 'arusah idini busah
  Ya 'aris 'albi baytub
Al-wad abu tabangah . . . al-busah minuh manga
Ya-halawah 'ala 'am' 'al-bamiyyah
  Umak tutbukh wa ukhtak tutbukh . . . wa anah 'alah higrak naymah
Ya-halawah 'ala 'am' 'al-kusah
  Umak tutbukh wa ukhtak tutbukh . . . wa anah 'alah 'ad al-busah

42. al-Sanfawi, *Aghani al-Hub wa al-Zawaj wa al-Afrah*, 179. A *qirat* is approximately 175 square meters.

Ummak bita'kilni al-mish . . . wannah mabahibish akl al-mish
'Alayah al-ni'ma labi' al-Tisht . . . wagib al-lahma al-dani ya wad
Ummak bithib al-Kurat . . . wannah mabahibish al-Kurat
'Alayah al-ni'ma labi' qirat . . . wagib al-lahma al-dani ya wad

43. Ahmad Taymur, *al-Amthal al-Amiyya* (Cairo: Dar al-Shuruq, 2010), 578, 237.

44. M. L. Whately, *Child-Life in Egypt* (Philadelphia: American Sunday School Union, 1866), 190. As related in Chapter 1, Whately was an English missionary and educator who in the 1860s and 1870s lived in the district of Bab al-Bahr in Cairo. For more on missionaries in nineteenth-century Egypt, see Paul Sedra, *From Mission to Modernity: Evangelicals, Reformers and Education in Nineteenth Century Egypt* (New York: I. B. Tauris, 2011).

45. Elizabeth Wickett, *For the Living and the Dead: The Funerary Laments of Upper Egypt, Ancient and Modern* (New York: I. B. Taurus, 2010), 101.

46. For the possible role of obituaries in class formation, see Hussein Omar, "'Snatched by Destiny's Hand': Obituaries and the Making of Class in Modern Egypt," *History Compass* 15, no. 6 (2017): 1–12.

47. For the importance of a better engagement with "death studies" in modern Middle East history, see Shane Minkin, "History from Six-feet Below: Death Studies and the Field of Modern Middle East History," *History Compass* 11, no. 8 (2013): 632–646.

48. Taymur, *al-Amthal al-Amiyya*, 549.

49. Early, *Baladi Women of Cairo*, 110.

50. Radwan, *Khati al-Ataba*, 59. Radwan connects this custom back to ancient Egypt.

51. Lewis Parkhurst, *A Vacation on the Nile: A Collection of Letters Written to Friends at Home* (Boston: Atheneum Press: 1913), 87. For more firsthand accounts of music in funerary processions, see Radwan, *Khati al-Ataba*, 59.

52. Sladen, *Oriental Cairo*, 223.

53. Edith Louisa Butcher, *Things Seen in Egypt* (London: Seeley Service, 1912 [1910]), 57–58.

54. Umran, *al-Ghina' al-Fulkluri* (Cairo: al-Markaz al-Misri Lil-Thaqafah wa al-Funun, 2012), 40; Sladen, *Oriental Cairo*, 220.

55. Radwan, *Khati al-Ataba*, 59.

56. Abd al-Halim Hifni, *al-Mirath al-Sha'bi: al-'Adid* (Cairo: al-Hay'a al-Misriyya al-'Ammah lil-Kitab, 1997), 116–117.

Ya binti ibki ala abuki . . . la yinfa'ik guzik wa la akhuki
Ma yi'iz al-wiliyya ghayr walidha . . . yis'al 'alayha wi yiqul nayibha
Ma yi'iz al-wiliyya ghayr abuha al-zayn . . . yis'al 'alayha wi yiqul rahit fayn
Ya-zayn abuyah, 'a'at fi-rihuh . . . yi'lam bi hali qabl ma nibihuh
Ya-zayn abuyah, 'a'at fi-dilluh . . . yi'lam bi hali qabl ma niquluh

57. Hifni, *al-Mirath al-Sha'bi'*, 18.

58. Wickett, *For the Living and the Dead*, 67.

59. Wickett, *For the Living and the Dead*, 11.

60. Hifni, *al-Mirath al-Sha'bi'*, 16–17.

6.1. For more on the dialogical elements of these mourning lamentations, see Wickett, *For the Living and the Dead*, 104.

62. Wickett, *For the Living and the Dead*, 122.

63. Wickett, *For the Living and the Dead*, 14–18.

64. For two variations of "The Stranger's Lament," see Hifni, *al-Mirath al-Sha'bi'*, 83–89; and Wickett, *For the Living and the Dead*, 20.

65. Wickett, *For the Living and the Dead*, 20.

66. For an excellent examination of the continuation between some of the mourning rituals of ancient and of modern Egypt, see Wickett, *For the Living and the Dead*.

67. Wickett, *For the Living and the Dead*, 113.

68. For example, see Baedeker, *Egypt and the Sudan*, xciv.
69. "Allah Hayy," "Ina lillahi wa ina illayhi raji'un," and "la hawla wa la quwata illa billah."
70. Radwan, *Khati al-Ataba*, 58–59.
71. Radwan, *Khati al-Ataba*, 58–59.
72. *Cook's Handbook for Egypt and the Egyptian Sudan*, 850.
73. Radwan, *Khati al-Ataba*, 58.
74. Radwan, *Khati al-Ataba*, 59.
75. Faris Khidr, *Mirath al-'Asa: Tasawwurat al-Mawt fi al-Wa'yi al-Sha'bi* (Cairo: al-Hayyi'a al-Misriyya al-'Amma lil-Kitab, 2009), 122–123.
76. Khidr, *Mirath al-'Asa*, 123.
77. Khidr, *Mirath al-'Asa*, 99.
78. Butcher, *Things Seen in Egypt*, 57–58.
79. For a description of a funerary *shadir* (*sirdaq*) in the Sayyida Zaynab district circa 1920, see Radwan, *Khati al-Ataba'*, 88.
80. Wickett, *For the Living and the Dead*, 70–71.
81. El-Sayed El-Aswad, "Death Rituals in Rural Egyptian Society: A Symbolic Study," *Urban Anthropology and Studies of Cultural Systems and World Economic Development* 16, no. 2 (1987): 230.
82. For a "modernist" critique of loud wailing at traditional Egyptian funerals, see Jirjis Antun, *al-'Insaniyyah wa-al-Tamadun* (Cairo: Matba'at al-Ma'arif, 1912), 79–80.
83. "Islah al-'Adat," *al-Muftah*, February 15, 1906.
84. The two-part editorial appeared in the February 15 and March 15, 1906, issues of *al-Muftah*. In the course of it, the writer describes the "civilized" manner in which a well-to-do French family celebrates the birth of a new child, and also describes an upper-class French wedding, complete with a detailed description of a waltz.
85. For a modernist critique of lavish and loud Egyptian weddings, see Antun, *al-'Insaniyyah wa-al-Tamadun*, 84.
86. Al-Muwaylihi's book first came out in serialized form in *Misbah al-Sharq* magazine, from 1898 to 1900. The first Arabic edition of the entire book was published by Matba'it al-Ma'arif in Cairo in 1907. Muhammad al-Muwaylihi, *What 'Isa ibn Hisham Told Us, or, a Period of Time*, trans. and ed. Roger Allen (New York: New York University Press, 2015 [1907]), 1: 331–343; Muhammad 'Umar, *Hadir al-Misriyyin 'Aw Sirr Ta'akhurihum* (Cairo: Matba'at al-Muqtataf, 1902), 10–13, 195–205, 270–275.
87. 'Umar, *Hadir al-Misriyyin 'Aw Sirr Ta'akhurihum*, 196.
88. 'Umar, *Hadir al-Misriyyin 'Aw Sirr Ta'akhurihum*, 197. According to Ahmad Amin, the Awlad Rabiyya players mentioned by 'Umar were a well-known and very popular street theater troupe hired for weddings and *mawalid*. They were known for their dirty humor and jokes. Ahmad Amin, *Qamuus al-'Adat wa al-Taqalid wa al-Ta'abir al-Misriyya* (Cairo: Dar al-Shuruq, 2010 [1953]), 18.
89. Muhammad Bakhit, *Kitab Ahsan al-Kalam fi ma Yata'allaqu bi al-Sunnah wa al-Bid'ah min al-Ahkam* (Cairo: Matba'at al-Kurdistan al-'Ilmiyah, 1911), 36–37.

90. Muhammad Bakhit, *Kitab Ahsan al-Kalam fi ma Yata'allaqu bi al-Sunnah wa al-Bid'ah min al-Ahkam*, 37–39.

91. Muhammad Bakhit, *Kitab Ahsan al-Kalam fi ma Yata'allaqu bi al-Sunnah wa al-Bid'ah min al-Ahkam*, 80.

92. Wickett, *For the Living and the Dead*, 113.

93. Khidr, *Mirath al-'Asa*, 99.

94. At the turn of the twenty-first century, so called Islamic weddings with varying degrees of music dancing, and entertainment were somewhat in vogue. See Karin van Nieuwkerk, *Performing Piety: Singers and Actors in Egypt's Islamic Revival* (Austin: University of Texas Press, 2013), 206–231.

Chapter 6

1. R. Murray Schafer, *The Soundscape: Our Sonic Environment and the Tuning of the World* (Rochester, NY: Destiny Books, 1994 [1977]), 76.

2. For an excellent analysis of these tensions as manifested in anti-*mawalid* sentiments, see Samuli Schielke, "Hegemonic Encounters: Criticism of Saints-Day Festivals and the Formation of Modern Islam in Late 19th and Early 20th-Century Egypt," *Die Welt des Islams* 47, no. 3/4 (2007): 319–355; and Meir Hatina, "Religious Culture Contested: The Sufi Ritual of Dawsa in Nineteenth-Century Cairo," *Die Welt des Islams* 47, no. 1 (2007): 33–62.

3. For an excellent examination of the building of Khedival power via aesthetics, architecture, and elite Arabic theater, see Adam Mestyan, *Arab Patriotism: The Ideology and Culture of Power in Late Ottoman Egypt* (Princeton; Princeton University Press, 2017).

4. *Al-Muqtataf*, April 1, 1895.

5. For an excellent analysis of timekeeping and the use of the citadel guns, see Daniel A. Stolz, "Positioning the Watch Hand: 'Ulama and the Practice of Mechanical Time Keeping in Cairo, 1737–1874," *International Journal of Middle East Studies* 47 (2015): 489–510.

6. This was also the case throughout the Ottoman Empire. See Adam Mestyan, "Upgrade? Power and Sound during Ramadan and *'Id al-Fitr* in the Nineteenth-Century Ottoman Arab Provinces," *Comparative Studies of South Asia, Africa and the Middle East* 37, no. 2 (2017): 267.

7. *Al-Muqtataf*, April 1, 1895.

8. *Egyptian Gazette*, March 17, 1924. The *Gazette*'s report covers the opening of parliament two days earlier, on March 15, 1924. A copy of the *Gazette* is included in the report from J. Morten Howell (U.S. Consul General) to U.S. Secretary of State (March 17, 1924), Records of the Department of State Relating to the Internal Affairs of Egypt, 1910–29 (National Archives Microfilm Publications, Microcopy no. 571, 883:032/5, Roll 9), National Archives at College Park, College Park, MD.

9. Report from J. Morten Howell (U.S. Consul General) to U.S. Secretary of State (November 13, 1924), Records of the Department of State Relating to the Internal Affairs

of Egypt, 1910–29. (National Archives Microfilm Publications, Microcopy no. 571, 883:032/9, Roll 9), National Archives at College Park, College Park, MD.

10. *Al-Bahlawan*, October 18, 1925. Technically, Fuad became Sultan of Egypt in October 1917 and King of Egypt after Egypt's nominal independence from Britain was achieved in March 1922.

11. Edith Louisa Butcher, *Things Seen in Egypt* (London: Seeley Service, 1912 [1920]), 159–160.

12. For determining the timing of Ramadan and the use of the Ramadan Cannon throughout the Arab provinces of the Ottoman Empire, see Mestyan, "Upgrade?"

13. *Al-Ithnayn wa al-Duniya*, May 10, 1954.

14. "Telegraph from the Shari'a Judge in Port Said to His Highness the Khedive" (in Arabic), DWQ, Majlis al-Wizara' Files, no. 0075-012985 (January 29, 1898).

15. "Telegraph from the Shari'a Judge in Port Said to His Highness the Khedive."

16. "Letter from Ministry of War to Council of Ministers" (in Arabic), 1914), DWQ, Majlis al-Wizara' Files, no. 0075-027534 (August 5).

17. "Letter from Ministry of War to Council of Ministers."

18. "Letter from Council of Ministers to Interior" (in Arabic), DWQ, Majlis al-Wizara' Files, no.0075-027534 (August 9, 1914).

19. *Al-Ahram*, January 17, 1934.

20. For a more contemporary look at Ramadan celebrations, see Walter Armbrust, "Synchronizing Watches: The State, the Consumer, and Sacred Time in Ramadan Television," in *Religion, Media, and the Public Sphere*, ed. Birgit Meyer and Annelies Moors (Bloomington: Indiana University Press, 2005), 207–226.

21. Frederic Courtland Penfield, *Present-Day Egypt* (New York: Century, 1903), 56–57.

22. For a brief history of military brass bands under Muhammad Ali in Egypt, see Adam Mestyan, "Sound, Military Music, and Opera in Egypt during the Rule of Mehmet Ali Pasha (r. 1805–1848)," in *Ottoman Empire and European Theatre*, vol. 2 of *The Time of Joseph Haydn—From Sultan Mahmud I to Mahmud II (r. 1730–1839)*, ed. Michael Huttler and Hans Ernst Weidinger (Vienna: Hollitzer Wissenschaftsverlag, 2014), 631–656.

23. Mestyan, *Arab Patriotism*, 86–89.

24. Alfred Cunningham, *Today in Egypt: Its Administration, People and Politics* (London: Hurst, 1912), 33.

25. Cunningham, *Today in Egypt*, 32.

26. The effect of Egyptian military musicians on the proliferation of classical Western music in Egypt is examined in Salwa el-Shawan, "Western Music and Its Practitioners in Egypt (ca. 1825–1985): The Integration of a New Musical Tradition in a Changing Environment," *Asian Music* 17, no. 1 (1985): 143–153.

27. Douglas Sladen, *Oriental Cairo: The City of the Arabian Nights* (Philadelphia: Lippincott, 1911), 258.

28. Thomas Harrison, *The Homely Diary of a Diplomat in the East, 1897–1899* (Boston, Houghton Mifflin, 1917), 101–103. "Hail Columbia" was one of the unofficial national

anthems for the United States, until "The Star Spangled Banner" became the official anthem in 1931.

29. U.S. Department of State Records of the Department of State Relating to the Internal Affairs of Egypt, 1910–29 (National Archives Microfilm Publications, Microcopy no. 571, 883:0159, Roll 8), National Archives at College Park, College Park, MD.

30. *Al-Radiu al-Misri*, March 21, 1936. "Id Milad Sahib al-'Arsh" literally translates to "the birthday of the owner of the throne."

31. *Al-Radiu al-Misri*, March 21, 1936. Ironically, Fuad would pass away less than two months after this issue appeared.

32. Ali Pasha Mubarak, *al-Khitat al-Tawfiqiyya al-Jadida li-Misr wa al-Qahira wa-Mudunaha wa-Biladaha al-Qadima wa al-Shahira* (Cairo: Matba'at Dar al-Kutub wa al-Watha'iq al-Qawmiyya, 2004 [1887/88]), 1: 226–232.

33. Butcher, *Things Seen in Egypt*, 159, 168.

34. By most accounts, most *mawalid* fit nicely with Mikhail Bakhtin's description of carnivals and the carnivalesque. Mikhail Bakhtin, *Problems of Dostoevsky's Poetics*, ed. and trans. Caryl Emerson (Minneapolis: University of Minnesota Press, 1984), 122–123.

35. See, for example, Schielke, "Hegemonic Encounters," 319–355; and Hatina, "Religious Culture Contested," 33–62.

36. Hatina, "Religious Culture Contested," 59.

37. See, for example, *al-Muqtataf*, March 1895 and October 1895; *al-Manar*, December 25, 1916; and Muhammad 'Umar, *Hadir al-Misriyyin 'Aw Sirr Ta'akhurihum* (Cairo: Matba'at al-Muqtataf, 1902), 262–265.

38. See, for example, Schielke, "Hegemonic Encounters," 319–355; and Hatina, "Religious Culture Contested," 33–62.

39. *Al-Ithnayn wa al-Duniya*, June 25, 1934.

40. Muhammad Bakhit, *Kitab Ahsan al-Kalam fi ma Yata'allaqu bi al-Sunnah wa al-Bid'ah min al-Ahkam* (Cairo: Matba'at al-Kurdistan al-'Ilmiyah, 1911), 73.

41. For example, see "Papers concerning Mawalid celebrations" (in Arabic), DWQ, Watha'iq 'Abdin Files, no. 0069-014985 (1915–1951).

42. For an interesting analysis of the ruling authorities and attempts at silencing the noises of festivals, see Jacques Attali, *Noise: The Political Economy of Music* (Minneapolis: University of Minnesota Press, 1985), 21–27.

43. Mikhail Bakhtin, *Rabelais and His World* (Bloomington: Indiana University Press, 1984), 9.

44. The word *mahmal* roughly translates as "that which is transported," and was the common term for the covered litter carrying people or goods on camelback.

45. Schafer, *Soundscape*, 76.

46. For more detail on the Mahmal, see On Barak, *On Time: Technology and Temporality in Modern Egypt* (Berkeley: University of California Press, 2013), 85–91.

47. Bakhit, *Kitab Ahsan al-Kalam fi ma Yata'allaqu bi al-Sunnah wa al-Bid'ah min al-Ahkam*, 77.

48. Bakhit, *Kitab Ahsan al-Kalam fi ma Yata'allaqu bi al-Sunnah wa al-Bid'ah min al-Ahkam*, 78.

49. Bakhit, *Kitab Ahsan al-Kalam Fima Yata'allaqu bi al-Sunnah wa al-Bid'ah min al-Ahkam*, 75.

50. Sladen, *Oriental Cairo*, 226.

51. Sladen, *Oriental Cairo*, 226–227.

52. For example, see "Papers concerning the buying of supplies and the expenses incurred for the celebration of the Prophet's Birthday Celebration" (in Arabic), DWQ, Watha'iq 'Abdin Files, no. 0069-020204 (1921–1949).

53. For example, see "Papers concerning Mawalid celebrations" (1915–1951).

54. Sladen, *Oriental Cairo*, 231.

55. Sladen, *Oriental Cairo*, 113.

56. Sladen, *Oriental Cairo*, 226–227.

57. Butcher, *Things Seen in Egypt*, 168.

58. Eustace Reynolds-Ball, *Cairo of Today: A Practical Guide to Cairo and the Nile* (London: A & C Black, 1916), 111–112; Sladen, *Oriental Cairo*, 239–240.

59. The picture in Figure 6.6 was taken on January 19, 1949. The back of the picture notes: "Birthday Parade: Cairo: on the Birthday of the Prophet. This colorful parade is part of the traditional celebrations. These regimented drummers march through al-Ghaffir Square which is bedecked with many marquees adding to the festivities. The crack Egyptian army band parades at the ceremony ground where Prime Minister Abdel Hadi represented His Majesty King Faruq (1/19/1949)."

60. Sladen, *Oriental Cairo*, 228.

61. For example, see Carolyn Birdsall, *Nazi Soundscapes: Sound, Technology and Urban Space in Germany, 1933–1945* (Amsterdam: Amsterdam University Press, 2012).

62. "Egypt: Succession to Throne" (May 4, 1936) and "Report from Lampson to Eden" (May 20, 1936), TNA, FO 141/757/1.

63. For a text of the King's Speech, see *Al-Waqa'i' al-Misriyya*, May 9 1936.

64. *Al-Radiu al-Misri*, May 16, 1936.

65. *Al-Musawwar*, May 30, 1938.

66. *Al-Radiu al-Misri*, (May 30, 1936).

67. "Reviving Ramadan Nights" (in Arabic), DWQ, Watha'iq 'Abdin Files, no. 0069-003402 (1946–1947). See also *Al-Musawwar*, May 30, 1938.

68. For more detail, see Mine Ener, *Managing Egypt's Poor and the Politics of Benevolence, 1800–1952* (Princeton: Princeton University Press, 2003), 24–26, 130, 139.

Conclusion

1. For example, see Peter Charles Hoffer, *Sensory Worlds in Early America* (Baltimore: Johns Hopkins University Press, 2003); Caroline A. Jones, *Eyesight Alone: Clement Greenberg's Modernism and the Bureaucratization of the Senses* (Chicago: University of Chicago Press, 2005); Roger Horowitz, *Putting Meat on the American Table: Taste, Technology, Transformation* (Baltimore: Johns Hopkins University Press, 2006); Mark M. Smith,

How Race Is Made: Slavery, Segregation, and the Senses (Chapel Hill: University of North Carolina Press, 2006); and Alain Corbin, *The Foul and the Fragrant: Odor and the French Social Imagination* (Cambridge, MA: Harvard University Press, 1988).

2. Samah Selim, *The Novel and the Rural Imaginary in Egypt, 1880–1985* (New York: Routledge, 2004), 58–60; Michael Ezekiel Gasper, *The Power of Representation: Publics, Peasants, and Islam in Egypt* (Stanford: Stanford University Press, 2009), 3, 207, 222; Lucie Ryzova, *The Age of the Efendiyya: Passages to Modernity in National Colonial Egypt* (Oxford: Oxford University Press, 2014), 150–151; Samuli Schielke, "Hegemonic Encounters: Criticism of Saints-Day Festivals and the Formation of Modern Islam in Late 19th and Early 20th-Century Egypt," *Die Welt des Islams* 47, no. 3/4 (2007): 346.

3. Ryzova, *Age of the Efendiyya*, 150–151.

4. In written nineteenth-century sources, *ruʿaʿ*, *sawqah*, and *ghawghaʾ*, all of which roughly translate to "the rabble," were commonly used to negatively classify Egypt's lower classes.

5. For much more on the positive and negative uses of the word *baladi*, see Evelyn Early, *Baladi Women of Cairo: Playing with an Egg and a Stone* (London: Lynne Reimer, 1993).

6. The case for the senses and sensory marginalization has been successfully used for studies examining race and racial construction in the U.S. south and elsewhere, though less so for class formation. See Mark M. Smith, *Listening to Nineteenth-Century America* (Chapel Hill: University of North Carolina Press, 2001); Smith, *How Race Is Made*; and Jennifer Lynn Stoever, *The Sonic Color Line: Race and the Cultural Politics of Listening* (New York: New York University Press, 2016).

7. A *niqrizan* is a one-sided drum, often hung on a string around the drummer's shoulders. Unlike the tabla (which is played by hand), the *niqrizan* is played with drumsticks.

8. Fathi Radwan, *Khati al-Ataba: Hayyat Tifl Misri* (Cairo: Dar al-Maʿarif, 1973), 90.

9. Most likely, *ta-ta* is derived from the ancient Egyptian word *titi*, which means "to walk" or "arrive."

10. Radwan, *Khati al-Ataba*, 126–129.

11. R. Murray Schafer, *The Soundscape: Our Sonic Environment and the Tuning of the World* (Rochester, NY: Destiny Books, 1994 [1977]), 180.

12. Unlike Egypt's print media, the music and film industries tend to be more positive in their outlook on Egypt's working class. Also, the Egyptian mass media industry has made a living on commemorating and remembering Egypt's past, including of course many of the lost sounds in soundscapes.

13. The first live performance of the song of the water sellers was in the play *Ululuh* by Naguib al-Rihani, in the summer of 1918. It was soon after recorded on disk by the Odeon Corporation. See Odeon #47645, "Lahn al-Saqayyin."

14. Darwish and Khayri were particularly adept at mimicking and amplifying the sounds and feelings of the Egyptian streets. See Ziad Fahmy, *Ordinary Egyptians: Creating the Modern Nation through Popular Culture* (Stanford: Stanford University Press, 2011).

15. For more analysis of this song and its lyrics, see Ziad Fahmy, "An Earwitness to History: Street Hawkers and Their Calls in Early 20th-Century ?Egypt," Roundtable, *International Journal of Middle East Studies* 48, no.1 (2016): 129–134.

16. For example, see "Shid al-Hizam" (Odeon #47658); "Hiz al-Hilal Ya Sayyid" (Odeon #47657); "Til'it ya Mahla Nurha" (Mechian #652); "Lahn al-Saqayyin" (Odeon #47645); "Ya Halwayit Um Isma'il" (Odeon #47608); "Ihna Ya Fandim Tugar al-'Agam" (Odeon #4652); "Lahn al-Siyas" (Odeon #47712); and "Basara Baragah" (Baydafone #82561). For the lyrics of most of these songs, see Maktabat al-Askandariyya, *Sayyid Darwish: Mawsu'at 'i'lam al-Musiqa al-'Arabiyya*, 2 vols. (Cairo: Dar al-Shuruq, 2003).

17. Salah Jahin became the premiere vernacular poet (*zaggal*) in Egypt during the second half of the twentieth century, inheriting that mantle from Bayram al-Tunsi.

18. For more on *al-Layla al-Kabira*, see Samuli Schielke, *The Perils of Joy: Contesting Mulid Festivals in Contemporary Egypt* (Syracuse: Syracuse University Press, 2012), 144–151; Samia Mehrez, "Representing the Moulid: Salah Jahin's Al-Layla al-Kabira between Populist and Nationalist Aspirations," in *Medieval and Early Modern Performance in the Eastern Mediterranean*, eds. Arzu Öztürkmen and Evelyn Birge Vitz (Turnhout: Brepols, 2014): 453–463.

19. For the full lyrics, see Salah Jahin, *Al-Layla al-Kabira wa Khamas Masrahiyat* (Cairo: Markaz al-Ahram lil Tarjama wa al-Nashr, 1992), 9–17.

20. Schielke, *Perils of Joy*, 151.

21. Schielke, *Perils of Joy*, 151; Mehrez, "Representing the Moulid," 462.

Bibliography

Archival Sources

Egypt

Dar al-Watha'iq al-Qawmiyya (DWQ) (Egyptian National Archives), Cairo
DWQ, 0075 Majlis al-Wizara' Files (Council of Ministers Files)
DWQ, 0069 Watha'iq 'Abdin (Abdin Palace Files)
DWQ, 2002 Dabtiyyat Misr (The Cairo Police)

Dar al-Kutub al-Misriyya (Egyptian National Library), Cairo
Arab language periodicals (1890–1955).

United Kingdom

The National Archives (TNA) of the United Kingdom, Kew, London
FO 141—Embassy and Consular Archives, Egypt, Correspondence
FO 371—General Correspondence, Political
FO 407—Confidential Print, Egypt and the Sudan
FO 953—Information Office
WO—War Office

United States of America

Records of the Department of State Relating to the Internal Affairs of Egypt, 1910–29 (National Archives Microfilm Publications, Microcopy no. 571, 883: 10–40, Rolls 1–15); National Archives at College Park, College Park, MD.

Published Government Sources and Indexes

Egypt

Annuaire statistique de l'Egypte [Statistical yearbook of Egypt for 1909, 1912, 1914, 1918, and 1928–29, 1935–1936, 1950, 1951–1954]. Ministère des finances, Départment de la statistique. Cairo: Imprimerie Nationale.

Bulis Misr, *Nizam al-Murur bil-Qahirah: Ta'limat li-hafz al-Amn al-'Am*. Cairo: Hikimdar Bulis Misr, 1938.

Dar al-Kutub al-Misriyya. *Fihris al-Dawriyat al-'Arabiyya Allati Taqtaniha al-Dar* [Index of the periodicals contained in the Egyptian National Library]. Cairo: Dar al-Kutub al-Misriyya, 1996.

Dar al-Kutub wa al-Watha'iq al-Qawmiyya. *Fihris al-Musiqa wa al-Ghina' al-'Arabi al-Qadim: al-Mussajala 'Ala 'Istiwanat* [Index of all the musical records contained in the Egyptian National Library]. Cairo: Matba'at Dar al-Kutub al-Misriyya, 1998.

Egypt. Tourism Ministry. *Al-Qahira: Kharitah Siyahiyya: al-Amakin al-Hamah wa Khutut al-Tram wa al-Utubis.* Cairo: Maslahit al-Siyaha, 1953.

Garstin, William. *Report upon the Administration of the Public Works Department in Egypt for 1902.* Cairo: National Printing Department, 1903.

Maktabat al-Askandariyya, *Sayyid Darwish: Mawsu'at 'I'lam al-Musiqa al-'Arabiyya.* 2 vols. Cairo: Dar al-Shuruq, 2003.

McLean, W. H. *Municipality of Alexandria: City of Alexandria Town Planning Scheme.* Cairo: Government Press, 1921.

Ministry of Public Works. *Annual Report for 1899.* Cairo: Government Press, 1900.

Ministry of Public Works. *Annual Report for 1921–1922.* Cairo: Government Press, 1925.

Ministry of the Interior. *Annual Report of the Cairo City Police for the Year 1937.* Cairo: Government Press in Bulaq, 1938.

Ministry of the Interior. *Annual Report of the Cairo City Police for the Year 1938.* Cairo: Government Press in Bulaq, 1939.

Ministry of the Interior. *Annual Report of the Cairo City Police for the Year 1940.* Cairo: Government Press in Bulaq, 1941.

Mubarak, Ali Pasha. *Al-Khitat al-Tawfiqiyya al-Jadida li Misr wa al-Qahira wa-Mudunaha wa-Biladaha al-Qadima wa al-Shahira.* Vols. 1–7. Cairo: Matba'at Dar al-Kutub wa al-Watha'iq al-Qawmiyya, 2004 [1887/88].

Nazzaarat al-Dakhiliyya: Qawanin wa-lawa'ih al-Bulis al-Misri. Cairo: al-Matba'a al-Kubra al-Amiriyya, 1895.

Report of the Ministry of Public Works for the Year 1912. Ministry of Public Works, Egypt. Cairo: Government Press, 1914.

Sami Pasha, Amin. *Taqwim al-Nil.* Vol. 3. Cairo: Matba'at Dar al-Kutub al-Misriyya, 1936.

Tarjamat La'iha: Tata'alaq bi Isti'mal al-Afrad al-Turuq al-'Umumiyya. Cairo: al-Matba'a al-Kubrah al-Amiriyya, 1903.

United Kingdom
Parliamentary Papers

Parliamentary Papers, House of Commons and Command. *Further Paper Respecting the Attack on British Officers at Denshawai* [Dinshaway]. Egypt nos. 3 and 4 (1906).

Parliamentary Papers. House of Commons. *Reports by his Majesty's Agent and Consul-General on the Finances, Administration, and Condition of Egypt and the Sudan.* [Annual reports from 1903 to 1914.] London: Harrison.

Government Reports

Egypt: Economic and Commercial Conditions in Egypt. Issued by the Board of Trade. [Reports for 1948 and 1952. Various authors.] London: His Majesty's Stationery Office.

Report on Economic and Commercial Conditions in Egypt. Issued by the Department of Overseas Trade. [Reports for 1933, 1935, 1937, and 1939. Various authors.] London: His Majesty's Stationery Office.

Report on the Economic and Financial Situation of Egypt. Issued by the Department of Overseas Trade. [Reports for 1923, 1924, 1925, 1927, and 1928. Various authors.] London: His Majesty's Stationary Office.

Recorded Music and Comedic Sketches

"Basara Baragah" (Baydafone #82561)
"Hiz al-Hilal Ya Sayyid" (Odeon #47657)
"Ihna Ya Fandim Tugar al-'Agam" (Odeon #4652)
"Lahn al-Saqayyin" (Odeon #47645)
"Lahn al-Siyas" (Odeon #47712)
"Shid al-Hizam" (Odeon #47658)
"Til'it ya Mahla Nurha" (Mechian #652)
"Ya Halwayit Um Isma'il" (Odeon #47608)

Film and Television

Bab al-Hadid [Cairo station]. Directed by Youssef Chahine. Cairo, 1958.

Cairo street scenes—outtakes. Fox Movietone News Story 2–18, 1928. [Originally filmed December 28, 1928.] http://mirc.sc.edu/islandora/object/usc%3A32904.

Faqrah min Shari' Muhammad 'Ali: Zaffit Mutahir fi al-Haraniyya. Burnamij al-Shari' al-Misri [Egyptian Television], 1980. https://www.youtube.com/watch?v=n5t3SJMWcI4.

Firqit Hasaballah. Al-Jazeera documentary (2017). https://www.youtube.com/watch?v=ZqkYTc_wYdQ.

Arabic Periodicals

(All periodicals were printed in Cairo.)
Al-Ahram
Al-Bahlawan (*Shibin al-Kum*)
Al-Dik
Al-Ithnayn
Al-Ithnayn wa al-Duniya
Al-Kashkul
Al-Kashkul al-Mussawwar
Al-Lata'if al-Mussawwara

Majallat Sarkis
Al-Manar
Al-Masrah
Misr
Al-Mu'ayid
Al-Muftah
Al-Muqattam
Al-Muqtataf
Al-Musawwar
Al-Radiu al-Misri
Ruz al-Yusuf
Al-Sabah
Al-Sayf
Al-Waqa'i' al-Misriyya
Al-Zuhur

English and French Periodicals
The Century Magazine (New York)
Epigram Magazine (London)
L'Illustration (Paris)
The New York Times (New York)
The Times (London)
The Washington Post (Washington, DC)

Published Diaries, Memoirs, Letters, and Speeches
'Abd al-Nur, Fakhri. *Mudhakirat Fakhri 'Abd al-Nur, Thawrat 1919: Dur Sa'd Zaghlul wa al-Wafd fi al-Haraka al-Wataniyya*. Cairo: Dar al-Shuruq, 1992.

Amin, Ahmad. *Hayati*. Cairo: Mataba'at al-Hay'a al-Misriyya al-'Amma lil-Kitab, 2003 [1950].

Amin, Galal. *Madha 'Alamatni al-Haiyyah: Sirah Dhatiyya*. Cairo: Dar al-Shuruq, 2007.

Amin, Mustafa. *Min Wahid li-'Ashara*. Cairo: al-Maktab al-Misri al-Hadith Lil-Taba'a wa al-Nashr, 1977.

Fahmi, 'Abd al-Rahman. *Mudhakirat 'Abd al-Rahman Fahmi: Yawmiyat Misr al-Siyasiyya*. Part 1. Cairo: al-Hay'a al-Misriyya al-'Amma lil-Kitab, 1988.

Gum'a, Muhammad Lutfi. *Shahid 'Ala al-'Asr: Mudhakirat Muhammad Lutfi Gum'a 1886–1937*. Cairo: al-Hay'a al-Misriyya al-'Amma lil-Kitab, 2000.

Khayri, Badi'. *Mudhakirat Badi' Khayri: Khamsah wa Arbi'un Sana 'ala Adwa' al-Masrah*. Beirut: Dar al-Thakafa, n.d.

Lutfi al-Sayyid, Ahmad. *Qisat Hayati*. Cairo: Dar al-Hilal, 1982.

Masabni, Badi'a, and Nazik Basilah, *Mudhakarat Badi'a Masabni*. Beirut: Dar Maktabit al-Hayah 1960.

Al-Naqash, Salim Khalil. *Misr li al-Misriyyin*. Vol. 4. Cairo: al-Hay'a al-Misriyya al-'Amma lil-Kitab, 1986 [1884].
Radwan, Fathi. *Khati al-Ataba: Hayyat Tifl Misri*. Cairo: Dar al-Ma'arif, 1973.
Al-Rafi'i, 'Abd al-Rahman. *Mudhakirati: 1889–1951*. Cairo: Dar al-Hilal, 1952.
Ramzi, Ibrahim. *Masrahuna Ayyam Zaman wa Tarikh al-Fananin al-Qudama'*. Cairo: Matba'at al-Salam, 1984.
Al-Rihani, Naguib. *Mudhakirat Naguib al-Rihani*. Cairo: Dar al-Hilal, 1959.
Sha'rawi, Huda. *Mudhakirat Ra'idat al-Mar'a al-'Arabiyya*. Cairo: Dar al-Hilal, 1981.
Sharubim, Mikha'il. *Al-Kafi fi Tarikh Misr al-Qadim wa al-Hadith*. Vol. 4. Cairo: Maktabat Matbuli, 2004 [1898–1900].
———. *Al-Kafi fi Tarikh Misr al-Qadim wa al-Hadith*. Vol. 5, bk. 3. Cairo: Matba'at Dar al-Kutub wa al-Watha'iq al-Qawmiyya bil-Qahira, 2003 [composed ca. 1910].
Shafiq Pasha, Ahmad. *Mudhakirati fi Nisf Qarn: Min Sanat 1873 ila Sanat 1923*. Vol. 2. Cairo: al-Hay'a al-Misriyya al-'Amma lil-Kitab, 1994.
———. *Mudhakirati fi Nisf Qarn*. Vol. 3. Cairo: al-Hay'a al-Misriyya al-'Amma lil- Kitab, 1998 [1934].

English and French Travel Literature and Memoirs

Baedeker, Karl. *Egypt and the Sudan: Handbook for Travelers*. Leipzig: 1914.
Balls, William Lawrence. *Egypt of the Egyptians*. London: I. Pitman, 1920.
Briggs, Martin S. *Through Egypt in War-Time*. London: Frederick Stokes, 1918.
Butcher, Edith Louisa. *Things Seen in Egypt*. London: Seeley Service, 1912 [1910].
Carrington, Richard. *The Tears of Isis: The Story of a New Journey from the Mouth to the Source of the Nile*. London: Chatto & Windus, 1959.
Charmes, Gabriel. *Five Months at Cairo and in Lower Egypt*. London: Bentley Press, 1883.
Cook's Handbook for Egypt and the Egyptian Sudan. London: Thomas Cook, 1921.
Cunningham, Alfred. *Today in Egypt: Its Administration, People and Politics*. London: Hurst, 1912.
Dinning, Hector. *By-ways on Service: Notes from an Australian Journal*. London: Constable, 1918.
———. *Nile to Aleppo, with the Light-Horse in the Middle-East*. New York: Macmillan, 1920.
Harrison, Thomas. *The Homely Diary of a Diplomat in the East, 1897–1899*. Boston: Houghton Mifflin, 1917.
Howell, Joseph Morton. *Egypt's Past, Present and Future*. Dayton: Service, 1929.
Lane-Poole, Stanley. *The Story of Cairo*. London: J. M. Dent, 1902.
Keown, A. W. *Forward with the Fifth: The Story of Five Years' War Service, Fifth Inf. Battalion, A.I.F*. Melbourne: Specialty Press, 1921.
Loewenbach, Lothaire. *Promenade Autour de l'Afrique, 1907*. Paris: Flammarion, 1908.
Parkhurst, Lewis. *A Vacation on the Nile: A Collection of Letters Written to Friends at Home*. Boston: Atheneum Press: 1913.

Reynolds-Ball, Eustace. *Cairo of Today: A Practical Guide to Cairo and the Nile*. London: A & C Black, 1916.
Sladen, Douglas. *Oriental Cairo: The City of the Arabian Nights*. Philadelphia: Lippincott, 1911.
———. *Queer Things About Egypt*. London: Hurst & Blackett, 1911.
Whately, M. L. *Child-Life in Egypt*. Philadelphia: American Sunday-School Union Press, 1866.
Woolson, Constance Fenimore. *Mentone, Cairo and Corfu*. New York: Harper & Brothers, 1896.

Arabic Secondary Sources

'Abd al-Fatah. Mahmud. *Al-Mas'ala al-Misriyya wa al-Wafd*. Cairo: n.p., 1921.
Ahmad, Abd al-Hakim Khalil al-Sayid. *'Adat wa Taqalid al-Zawaj: Dirasah fi al-Thaqafah al-Sha'biyyah*. Cairo: Misr al-'Arabiyya lil-Dirasah wa al-Tawdhi, 2014.
———. *Dirasat fi al-Mu'taqad al-Sha'bi*. Cairo: al-Hay'ah al- 'Amma li-Qusur al-Thaqafah, 2000
al-'Allati, Hasan. *Kitab Tarwih al-Nufus wa Mudhik al-'Ubus*. 2 vols. Cairo: Matba'at Jaridat al-Mahrusa, 1889.
Amin, Ahmad. *Qamuus al-'Adat wa al-Taqalid wa al-Ta'abir al-Misriyya*. Cairo: Dar al-Shuruq, 2010 [1953].
Antun, Jirjis. *Al-'Insaniyyah wa-al-Tamadun*. Cairo: Matba'at al-Ma'arif, 1912.
al-'Azab, Yusri. *'Azjal Bayram al-Tunsi: Dirasa Faniyya*. Cairo: al-Hay'a al-Misriyya al-'Amma lil-Kitab, 1981.
Bakhit, Muhammad. *Kitab Ahsan al-Kalam fi ma Yata'allaqu bi al-Sunnah wa al-Bid'ah min al-Ahkam*. Cairo: Matba'at al-Kurdistan al- 'Ilmiyya, 1911.
Darwish, Hasan. *Min 'Ajl 'Abi: Sayyid Darwish*. Cairo: al-Hay'a al-Misriyya al-'Amma lil-Kitab, 1990.
Ghanim, Muhammad. *Afrah al-Ghalabah wa al-Akabir*. Cairo: Maktabit al-Anglo al-Misriyya, 2007.
Ghunaym, Muhammad. *Al-'Adat wa al-Taqalid fi Dilta Misr*. Cairo: 'Ayn lil-Dirasat wa al-Buhuth al-Insaniyya wa al-Ijtima'iyya, 2005.
al-Hadari, Ahmad. *Tarikh al-Sinima fi Misr: al-Guz' al-Thani, 1931–1940*. Cairo: al-Hay'a al-Misriyya al- 'Ammah lil-Kitab, 1989.
Haykal, Muhammad Husayn. *Zaynab*. Cairo: Maktabat al-Nahda al-Misriyya, 1967.
Hifni, Abd al-Halim. *Al-Mirath al-Sha'bi: al-'Adid*. Cairo: al-Hay'a al-Misriyya al-'Ammah lil-Kitab, 1997.
Jahin, Salah. *Al-Layla al-Kabira wa Khamas Masrahiyat*. Cairo: Markaz al-Ahram lil Tarjama wa al-Nashr, 1992.
Kamal, Safwat. *Al-Ghina' al-Sha'bi al-Misri: Mawawil wa Qisas Ghina'iyya Sha'biyya*. Cairo: al-Hay'a al-Misriyya al-'Amma lil-Kitab, 1994.
Kamil, Najwa. *Al-Sahafa al-Wafdiyya wa al-Qadiyya al-Wataniyya, 1919–1936*. Cairo: al-Hay'a al-Misriyya al-'Amma lil-Kitab, 1989.

Khidr, Faris. *Mirath al-'Asa: Tasawwurat al-Mawt fi al-Wa'yi al-Sha'bi.* Cairo: al-Hayyi'a al-Misriyya al- 'Amma lil-Kitab, 2009.

Kilani, Muhamad Sayyid. *Tram al-Qahirah: Dirasah Tarikhiyya, Ijtima'iyya, Adabiyya.* Cairo: Matba'at al-Madani, 1968.

Mahallawi, Hanafi. *Layali al-Qahirah fi 'Asr Badi'a: al-Raqisah Allati Hakamat Bashawat Misr wa-Sa'aliq 'Imad al-Din.* Cairo: Dar al-Kitab al- 'Arabi, 1997.

Mursi, Ahmad. *Al-Adab al-Sha'bi wa thakafat al-Mugtama.* Cairo: Dar Misr al-Mahrusa, 2008.

Nussayr, 'Ayda Ibrahim. *Al-Kutub al-'Arabiyya allati Nusharat fi Misr bayn 'Am 1900–1925.* Cairo: Qism al-Nashr bil-Jami'a al-Amrikiyya bil-Qahira, 1983.

———. *Al-Kutub al-'Arabiyya allati Nusharat fi Misr fi al-Qarn al-Tasi' 'Ashr.* Cairo: Qism al-Nashr bil-Jami'a al-Amrikiyya bil-Qahira, 1990.

———. *Harakat Nashr al-Kutub fi Misr fi al-Qarn al-Tasi' 'Ashr.* Cairo: al-Hay'a al-Misriyya al-'Amma lil-Kitab, 1994.

Rachid, Bahija. *Al-Taqatiq al-Sha'biyya.* Cairo: al-Lajna al-Musiqiyya al-'Ulya, 1968.

al-Rafi'i, 'Abd al-Rahman. *Al-Thawra al-'Urabiyya wal-Ihtilal al Injilizi,* 2 vols. Cairo, 1949.

———. *'Asr Isma'il.* Vol. 1. Cairo: Dar al-Ma'arif, 1987.

———. *Misr Wa al-Sudan fi 'Awa'il 'ahd al-Ihtilal, Tarikh Misr al-Qawmi Min Sanat 1882 'Ilá Sanat 1892.* Cairo: al-Dar al-Qawmiyya lil-Taba'a wa-al-Nashr, 1966.

al-Ra'i, Ali. *Funun al-Kumidiyya: Min Khayal al-Zil ila Naguib al-Rihani.* Cairo: Dar al-Hilal, 1971.

———. *Al-Kumidiyya al-Murtajila fi al Masrah al-Misri.* Cairo: Dar al-Hilal, 1968.

Ramadan, 'Abd al-'Azim. *Tatawwur al-Haraka al-Wataniyya fi Misr, 1918–1936.* Cairo: al-Hay'a al-Misriyya al-'Amma lil-Kitab, 1998.

Riyad, Husayn Mazlum, and Mustafa Muhammad al-Sabah. *Tarikh Adab al-Sha'b: Nash'atu, Tatawiratu, A'lamu.* Cairo: Matba'at al-Sa'ada, 1936.

Salih, Ahmad Rushdi. *Al-Adab al-Sha'bi.* Cairo: al-Hay'a al-Misriyya al-'Amma lil-Kitab, 2002 [1956].

Salim, Latifah Muhammad. *Misr fi al-Harb al-'Alamiyya al-'Ula.* Cairo: al-Hay'a al-Misriyya al-'Amma lil-Kitab, 1984.

al-Sanfawi, Fathi. *Aghani al-Hub wa al-Zawaj wa al-Afrah.* Cairo: Starburs lil-Tiba'a wa al-Nashr, 2009.

Shabana, Muhammad. "Aghani al-Zawaj fi Ba'd Qurah al-Mahalah al-Kubrah." In *Al-A'ras al-Sha'bia.* Cairo: Dar al-Za'im lil-Tiba'ah al-Hadithah, 2005.

Shadi, Ali Abu. *Waqa'i' al-Sinima al-Misriyya.* Cairo: al-Hayyi'a al-Misriyya al-'Amma lil-Kitab, 2003.

Taymur, Ahmad. *Al-Amthal al-Amiyya.* Cairo: Dar al-Shuruq, 2010.

'Umar, Muhammad. *Hadir al-Misriyyin 'Aw Sirr Ta'akhurihum.* Cairo: Matba'at al-Muqtataf, 1902.

Umran, Muhammad. *Al-Ghina' al-Fulkluri.* Cairo: al-Markaz al-Misri Lil-Thakafah wa al-Funun, 2012.

Zaghlul, Ahmad Fathi. *Sirr Taqaddum al-Injliz al-Saksuniyyin.* Cairo: Matbaʿat al-Maʿarif, 1899).
Zaki, ʿAbd al-Hamid Tawfiq. *Al-Tadhawuq al-Musiqi wa Tarikh al-Musiqa al-Misriyya.* Cairo: al-Hayʾa al-Misriyya al-ʿAmma lil-Kutub, 1995.

English and French Secondary Sources

Abu-Lughod, Lila. "Bedouins, Cassettes and Technologies of Public Culture." *Middle East Report,* no. 159 (July-August 1989): 25–30.

———. *Dramas of Nationhood: The Politics of Television in Egypt.* Chicago: University of Chicago Press, 2005.

———. "Writing against Culture." In *Recapturing Anthropology: Working in the Present,* edited by Richard G. Fox. Santa Fe, NM: School of American Research Press, 1991.

Ahmed, Jamal Mohammed. *Intellectual Origins of Egyptian Nationalism.* London: Oxford University Press, 1960.

Ahmed, Leila. *A Quiet Revolution: The Veil's Resurgence, from the Middle East to America.* New Haven: Yale University Press, 2012.

———. *Women and Gender in Islam: Historical Roots of a Modern Debate.* New Haven: Yale University Press, 1993.

Alsayyad, Nezar. "'Ali Mubarak's Cairo: between the Testimony of ʿAlamuddin and the Imaginary of the Khitat." In *Making Cairo Medieval,* edited by Nezar Alsayyad, Irene Bierman, and Nasser Rabbat, 49–68. Lanham, MD: Lexington Books, 2005.

Alsayyad, Nezar, Irene Bierman, and Nasser Rabbat, eds. *Making Cairo Medieval.* Lanham, MD: Lexington Books, 2005.

Anderson, Benedict. *Imagined Communities: Reflections on the Origins and Spread of Nationalism.* 2nd ed. New York: Verso Books, 1991.

Anderson, Kyle. "The Egyptian Labor Corps: Logistical Laborers in War and Revolution, 1914–1919." PhD diss., Cornell University, 2017.

Armbrust, Walter. "Audiovisual Media and History of the Middle East." In *History and Historiographies of the Modern Middle East,* edited by Amy Singer and Israel Gershoni, 288–312. Seattle: University of Washington Press, 2006.

———. "The Formation of National Culture in Egypt in the Interwar Period: Cultural Trajectories." *History Compass* 7, no. 1 (2009): 155–180.

———. *Mass Culture and Modernism in Egypt.* Cambridge: Cambridge University Press, 1996.

———. *Mass Mediations: New Approaches to Popular Culture in the Middle East and Beyond.* Berkeley: University of California Press, 2000.

———. "Synchronizing Watches: The State, the Consumer, and Sacred Time in Ramadan Television." In *Religion, Media, and the Public Sphere,* edited by Birgit Meyer and Annelies Moors, 207–226. Bloomington: Indiana University Press, 2005.

Attali, Jacques. *Noise: The Political Economy of Music.* Minneapolis: University of Minnesota Press, 1986.

Ayalon, Ami. *The Press in the Arab Middle East: A History*. London: Oxford University Press, 1995.
Badran, Margot. *Feminists, Islam, and Nation: Gender and the Making of Modern Egypt*. Princeton: Princeton University Press, 1995.
Badrawi, Malak. *Political Violence in Egypt 1910–1924: Secret Societies, Plots and Assassinations*. London: Curzon Press, 2000.
Baer, Gabriel. *Egyptian Guilds in Modern Times*. Jerusalem: Israel Oriental Society, 1964.
Bakhtin, Mikhail. *The Dialogic Imagination: Four Essays*. Austin: University of Texas Press, 2006.
———. *Problems of Dostoevsky's Poetics*. Edited and translated by Caryl Emerson. Minneapolis: University of Minnesota Press, 1984.
———. *Rabelais and His World*. Bloomington: Indiana University Press, 1984.
———. *Speech Genres and Other Late Essays*. Austin: University of Texas Press, 1986.
Barak, On. *On Time: Technology and Temporality in Modern Egypt*. Berkeley: University of California Press, 2013.
———. "Scraping the Surface: The Techno-Politics of Modern Streets in Turn-of-Twentieth-Century Alexandria." *Mediterranean Historical Review* 24, no. 2 (2009): 187–205.
Baron, Beth. *The Orphan Scandal: Christian Missionaries and the Rise of the Muslim Brotherhood*. Stanford: Stanford University Press, 2014.
———. *The Women's Awakening in Egypt: Culture, Society, and the Press*. New Haven: Yale University Press, 1994.
———. *Egypt as a Woman: Nationalism, Gender, and Politics*. Berkeley: University of California Press, 2005.
Beinin, Joel. "Writing Class: Workers and Modern Egyptian Colloquial Poetry (Zajal)." *Poetics Today* 15, no. 2 (Summer 1994): 191–215.
Beinin, Joel, and Zachary Lockman. *Workers on the Nile: Nationalism, Communism, Islam, and the Egyptian Working Class, 1882–1954*. Cairo: American University in Cairo Press, 1998.
Berlant, Lauren. *The Female Complaint: The Unfinished Business of Sentimentality in American Culture*. Durham, NC: Duke University Press, 2008.
Berque, Jacques. *Egypt: Imperialism and Revolution*. Translated by Jean Stewart. London: Faber and Faber, 1972.
Biancani, Francesca. "International Migration and Sex Work in Early Twentieth-Century Cairo." In *A Global Middle East: Mobility, Materiality and Culture in the Modern Age, 1880–1940*, edited by Liat Kozma, Cyrus Schayech, and Avner Wishnitzer, 109–133. New York: I. B. Tauris, 2014.
Birdsall, Carolyn. *Nazi Soundscapes: Sound, Technology and Urban Space in Germany, 1933–1945*. Amsterdam: Amsterdam University Press, 2012.
Booth, Marilyn. *Bayram al-Tunsi's Egypt: Social Criticism and Narrative Strategies*. Exeter: Ithaca Press, 1990.

———. "Colloquial Arabic Poetry, Politics, and the Press in Modern Egypt." *International Journal of Middle East Studies* 24, no. 3 (1992): 419–440.

———. *May Her Likes Be Multiplied: Biography and Gender Politics in Egypt*. Berkeley: University of California Press, 2001.

Botman, Selma. *Engendering Citizenship in Egypt*. New York: Columbia University Press, 1999.

Bourdieu, Pierre. *Distinction: A Social Critique of the Judgment of Taste*. Cambridge, MA: Harvard University Press, 2000.

Brass, Paul R. *Ethnicity and Nationalism: Theory and Comparison*. New Delhi: Sage. 1991.

Brown, Nathan J. *Peasant Politics in Modern Egypt: Struggle against the State*. New Haven: Yale University Press, 1990.

Bull, Michael, Paul Gilroy, David Howes, and Douglas Kahn. "Introducing Sensory Studies." *The Senses and Society* 1 (2006): 5–7.

Burckhardt, John Lewis. *Arabic Proverbs and the Manners and Customs of Modern Egyptians*. London: Bernard Quaritch, 1875.

Burke, Peter. *Popular Culture in Early Modern Europe*. London: Maurice Temple Smith, 1978.

Cachia, Pierre. *Popular Narrative Ballads of Modern Egypt*. Oxford: Clarendon Press, 1989.

———. "The Use of the Colloquial in Modern Arabic Literature." *Journal of American Oriental Studies* 87, no. 1 (January–March 1967): 12–22.

Carlier, Omar. "Le café maure: Sociabilité masculine et effervescence citoyenne (Algérie XVIIe–XXe siècles)." *Annales: Histoire, Sciences Sociales* 45, no. 4 (1990): 975–1004.

Carminati, Lucia. "Bū Saʿī/Port Said, 1859–1914: Migration, Urbanization, and Empire in an Egyptian and Mediterranean Port-City." PhD diss., University of Arizona, 2018.

Cengiz, Kirli "The Struggle over Space: Coffeehouses of Ottoman Istanbul, 1780–1845." PhD diss., State University of New York at Binghamton, 2000.

Cinar, Alev. *Modernity, Islam, and Secularism in Turkey: Bodies, Places, and Time*. Minneapolis: University of Minnesota Press, 2005.

Chalcraft, John T. *The Striking Cabbies of Cairo and Other Stories: Crafts and Guilds in Egypt, 1863–1914*. Albany: State University of New York Press, 2004.

Clifford, James. *The Predicament of Culture: Twentieth-Century Ethnography, Literature, and Art*. Cambridge, MA: Harvard University Press, 1988.

Cobb, Elvan. "Railway Crossings: Encounters in Ottoman Lands." PhD diss., Cornell University, 2018.

Cole, Juan R. I. *Colonialism and Revolution in the Middle East: Social and Cultural Origins of Egypt's ʿUrabi Movement*. Cairo: American University in Cairo Press, 1999.

Connor, Steven. "Edison's Teeth: Touching Hearing." In *Hearing Cultures: Essays on Sound, Listening and Modernity*, edited by Veit Erlmann, 153–172. New York: Berg, 2004.

Corbin, Alain. *The Foul and the Fragrant: Odor and the French Social Imagination*. Cambrige, MA: Harvard University Press, 1988.

———. *Time, Desire and Horror: Towards a History of the Senses*. Cambridge: Polity Press, 1995.

———. *Village Bells: Sound and Meaning in the Nineteenth-Century French Countryside*. New York: Columbia University Press, 1998.

Cuno, Kenneth M. *Modernizing Marriage: Family, Ideology, and Law in Nineteenth- and Early Twentieth-Century Egypt*. Syracuse: Syracuse University Press, 2015.

———. *The Pasha's Peasants: Land, Society, and Economy in Lower Egypt, 1740–1858*. Cambridge: Cambridge University Press, 1992.

Danielson, Virginia. "Artists and Entrepreneurs: Female Singers in Cairo during the 1920s." In *Women in Middle Eastern History: Shifting Boundaries in Sex and Gender*, edited by Nikki Keddie and Beth Baron, 292–309. New Haven: Yale University Press, 1991.

Davis, Eric. *Challenging Colonialism: Bank Misr and Egyptian Industrialization, 1920–1941*. Princeton: Princeton University Press, 1983.

De Certeau, Michel. *The Practice of Everyday Life*. Berkeley: University of California Press, 2011.

Deeb, Maurius. *Party Politics in Egypt: The Wafd and Its Rivals, 1919–1939*. London: Ithaca Press, 1979.

Demolins, Edmond. *A quoi tient la supériorité des Anglo-Saxons*. Paris: Firmin-Didot, 1898.

Derr, Jennifer. *The Lived Nile: Environment, Disease, and Material Colonial Economy in Egypt*. Stanford: Stanford University Press, 2019.

Desart, Roland. "The Cairo Tramways: Africa's Greatest Tramway Network," Part 1. *Modern Tramway and Light Rail Transit* (May 1991): 160–169.

———. "The Cairo Tramways: Africa's Greatest Tramway Network," Part 2. *Modern Tramway and Light Rail Transit* (June 1991): 193–205.

Dougherty, Roberta L. "Badi'a Masabni, Artiste and Modernist: The Egyptian Print Media's Carnival of National Identity." In *Mass Mediations: New Approaches to Popular Culture in the Middle East and Beyond*, edited by Walter Armbrust, 243–268. Berkeley: University of California Press, 2000.

Durrell, Lawrence. *Justine*. New York: Penguin Books, 1985.

Early, Evelyn. *Baladi Women of Cairo: Playing with an Egg and a Stone*. London: Lynne Reimer, 1993.

Eidsheim, Nina Sun. *The Race of Sound: Listening, Timbre, and Vocality in African American Music*. Durham, NC: Duke University Press, 2019.

El-Aswad, El-Sayed. "Death Rituals in Rural Egyptian Society: A Symbolic Study." *Urban Anthropology and Studies of Cultural Systems and World Economic Development* 16, no. 2 (1987): 230–250.

Ellis, Matthew. *Desert Borderland: The Making of Modern Egypt and Libya*. Stanford: Stanford University Press, 2019.

Elsaket, Ifdal. "Sound and Desire: Race, Gender, and Insult in Egypt's First Talkie." *International Journal of Middle East Studies* 51, no. 2 (2019): 203–232.

Elshahed, Mohamed. "Egypt Here and There: The Architectures and Images of National Exhibitions and Pavilions, 1926–1964." *Annales Islamologiques* 50 (2016): 107–143.

El-Shawan, Salwa. "Western Music and Its Practitioners in Egypt (ca. 1825–1950): The Integration of a New Musical Tradition in a Changing Environment." *Asian Music* 17 (1985): 143–153.

Ener, Mine. *Managing Egypt's Poor and the Politics of Benevolence, 1800–1952*. Princeton: Princeton University Press, 2003.

Erlmann, Veit, ed. *Hearing Cultures: Essays on Sound, Listening and Modernity*. New York: Berg, 2004.

Fahmy, Khaled. *All the Pasha's Men: Mehmed Ali, His Army and the Making of Modern Egypt*. Cambridge: Cambridge University Press, 1997.

———. *In Quest of Justice: Islamic Law and Forensic Medicine in Modern Egypt*. Berkeley: University of California Press, 2018.

———. "Modernizing Cairo: A Revisionist Narrative." In *Making Cairo Medieval*, edited by Nezar Alsayyad, Irene Bierman, and Nasser Rabbat, 173–200. Latham: Lexington Books, 2005.

———. "An Olfactory Tale of Two Cities: Cairo in the Nineteenth Century." In *Historians in Cairo: Essays in Honor of George Scanlon*, edited by Jill Edwards, 155–187. Cairo: American University in Cairo Press, 2002.

Fahmy, Ziad. "An Earwitness to History: Street Hawkers and their Calls in Early 20th-Century Egypt." Roundtable. *International Journal of Middle East Studies* 48, no.1 (2016): 129–134.

———. "Coming to Our Senses: Historicizing Sound and Noise in the Middle East." *History Compass* 11, no. 4 (April 2013): 305–315.

———. "Francophone Egyptian Nationalists, Anti-British Discourse, and European Public Opinion 1885–1910: The Case of Mustafa Kamil and Ya'qub Sannu'." *Comparative Studies of South Asia, Africa and the Middle East* 28, no. 1 (2008): 176–77.

———. "Jurisdictional Borderlands: Extraterritoriality and 'Legal Chameleons' in Precolonial Alexandria, 1840–1870." *Comparative Studies in Society and History* 55, no. 2 (April 2013): 305–329.

———. *Ordinary Egyptians: Creating the Modern Nation through Popular Culture*. Stanford: Stanford University Press, 2011.

Foucault, Michel. *The Birth of the Clinic: An Archaeology of Medical Perception*. NewYork: Random House Vintage Books, 1973.

———. "Interview by Sylvere Lotringer." In *Foucault Live: Interviews, 1966–84*, edited by Sylvere Lotringer. New York: Semiotex, 1989.

Fyfe, H. Hamilton. *The New Spirit in Egypt*. London: William Blackwood, 1911.

Gasper, Michael Ezekiel. *The Power of Representation: Publics, Peasants, and Islam in Egypt*. Stanford: Stanford University Press, 2009.

Gershoni, Israel, and James P. Jankowski. *Confronting Fascism in Egypt: Dictatorship Versus Democracy in the 1930s*. Stanford: Stanford University Press, 2010.

———. *Egypt, Islam and the Arabs: The Search for Egyptian Nationhood, 1900-1930*. New York: Oxford University Press, 1986.

———. "Print Culture, Social Change, and the Process of Redefining Imagined Communities in Egypt," *International Journal of Middle East Studies* 31, no. 1 (1999): 81-94.

———. *Redefining the Egyptian Nation, 1930-1945*. New York: Cambridge University Press, 1995.

Ghali, Ibrahim. *L'Égypte nationaliste et libérale de Moustapha Kamel à Saad Zaggloul, 1892-1927*. The Hague: Martinus Nijhoff, 1969.

Goldberg, Ellis. "Peasants in Revolt—Egypt 1919." *International Journal of Middle East Studies* 24, no. 2 (1992): 261-280.

Goldschmidt, Arthur, Jr. *Biographical Dictionary of Modern Egypt*. London: Lynne Reinner, 2000.

———. "The Egyptian Nationalist Party, 1892-1919." In *Political and Social Change in Modern Egypt*, edited by P. M. Holt, 308-333. London: Oxford University Press, 1968.

———. *Modern Egypt: The Formation of a State*. Boulder: Westview Press, 1988.

Gordon, Joel. "Broken Heart of the City: Youssef Chahine's *Bab al-Hadid* (Cairo Station)." *Journal for Cultural Research* 16, no. 2-3 (2012): 217-237.

———. *Nasser's Blessed Movement: Egypt's Free Officers and the July Revolution*. Oxford: Oxford University Press, 1992.

Gramsci, Antonio. *Prison Notebooks*. Translated by Joseph A. Buttigieg and Antonio Callari. New York: Columbia University Press, 1992.

Gran, Peter. *Islamic Roots of Capitalism: Egypt, 1760-1840*. Syracuse: Syracuse University Press, 1998.

———. "Upper Egypt in Modern History: A 'Southern Question'?" In *Upper Egypt: Identity and Change*, edited by Nicholas Hopkins and Reem Saad, 79-96. Cairo: American University in Cairo Press, 2004.

Gronow, Pekka. "The Record Industry Comes to the Orient." *Ethnomusicology* 25, no. 2 (1981): 251-284.

Haeri, Niloofar. *Sacred Language, Ordinary People: Dilemmas of Culture and Politics in Egypt*. New York: Palgrave Macmillan, 2003.

Hammad, Hanan. *Industrial Sexuality: Gender, Urbanization, and Social Transformation in Egypt*. Austin: Texas University Press, 2016.

Hanley, Will. *Identifying with Nationality: Europeans, Ottomans, and Egyptians in Alexandria*. New York: Columbia University Press, 2017.

Hanna, Nelly. *In Praise of Books: A Cultural History of Cairo's Middle Class, Sixteenth to the Eighteenth Century*. Syracuse: Syracuse University Press, 2003.

Hatina, Meir. "Religious Culture Contested: The Sufi Ritual of Dawsa in Nineteenth-Century Cairo." *Die Welt des Islams* 47, no. 1 (2007): 33-62.

Heyworth-Dunne, James. "A Selection of Cairo's Street Cries (Referring to Vegetables, Fruit, Flowers and Food)." *Bulletin of the School of Oriental Studies* 9 (1938): 350–359.

Hirschkind, Charles. *The Ethical Soundscape: Cassette Sermons and Islamic Counterpublics*. New York: Columbia University Press, 2006.

Hoffer, Peter Charles. *Sensory Worlds in Early America*. Baltimore: Johns Hopkins University Press, 2003.

Hogan, C. M. "Analysis of Highway Noise," *Journal of Water, Air, & Soil Pollution* 2, no. 3 (1973): 387–392.

Holt, P. M., ed. *Political and Social Change in Modern Egypt*. London: Oxford University Press, 1968.

Hopkins, Nicholas, and Reem Saad, eds. *Upper Egypt: Identity and Change*. Cairo: American University in Cairo Press, 2004.

Horowitz, Roger. *Putting Meat on the American Table: Taste, Technology, Transformation*. Baltimore: Johns Hopkins University Press, 2006.

Howes, David. *The Sixth Sense Reader*. London: Berg, 2009.

Hunter, F. Robert. *Egypt Under the Khedives 1805–1879: From Household Government to Modern Bureaucracy*. Cairo: American University in Cairo Press, 1999.

Ilber, Robert. *Heliopolis: Le Caire, 1905–1922*. Paris: Centre National de la Recherche-Scientifique, 1981.

Ismail, Salwa. *Political Life in Cairo's New Quarters: Encountering the Everyday State*. Minneapolis: University of Minnesota Press, 2006.

Jacob, Wilson Chacko. *Working Out Egypt: Effendi Masculinity and Subject Formation in Colonial Modernity, 1870–1940*. Durham, NC: Duke University Press, 2011.

Jacobs, Jennifer E. "Ululation in Levantine Society: The Cultural Reproduction of an Affective Vocalization." PhD Diss., University of Pennsylvania, 2008.

Jakes, Aaron George. "The Scales of Public Utility: Agricultural Roads and State Space in the Era of the British Occupation." In *The Long 1890s in Egypt: Colonial Quiescence, Subterranean Resistance*, edited by Marilyn Booth and Anthony Gorman, 57–86. Edinburgh: Edinburgh University Press, 2014.

Jankowski, James, and Israel Gershoni, eds. *Rethinking Nationalism in the Arab Middle East*. New York: Columbia University Press, 1997.

Jones, Caroline A. *Eyesight Alone: Clement Greenberg's Modernism and the Bureaucratization of the Senses*. Chicago: University of Chicago Press, 2005.

Kahn, Douglas. *Noise, Water, Meat: A History of Sound in the Arts*. Cambridge, MA: MIT Press, 1999.

Kapchan, Deborah, ed. *Theorizing Sound Writing*. Middletown, CT: Wesleyan University Press, 2017.

Kerr, Malcolm. *Islamic Reform: The Political Theories of Muhammad 'Abduh and Rashid Rida*. Los Angeles: University of California Press, 1966.

Khan, Noor. *Egyptian-Indian Nationalist Collaboration and the British Empire*. New York: Palgrave Macmillan, 2011.

Kholoussy, Hanan. *For Better, for Worse: The Marriage Crisis That Made Modern Egypt*. Stanford: Stanford University Press, 2010.

Khuri-Makdisi, Ilham. *The Eastern Mediterranean and the Making of Global Radicalism, 1860–1914*. Berkeley: University of California Press, 2013.

Kozma, Liat. *Policing Egyptian Women: Sex, Law, and Medicine in Khedival Egypt*. Syracuse: Syracuse University Press, 2011.

Lagrange, Frédéric. "Musiciens et poètes en Égypte au temps de la nahda." PhD diss., Université de Paris à Saint-Denis, 1994.

Landau, Jacob. *Studies in the Arab Theater and Cinema*. Philadelphia: University of Pennsylvania Press, 1969.

Lane, Edward William. *An Account of the Manners and Customs of the Modern Egyptians: Written in Egypt during the Years 1833–1835*. London: East West, 1978.

Lefebvre, Henri. *The Production of Space*. Malden: Blackwell, 1991.

Lockman, Zachary. "Imagining the Working Class: Culture, Nationalism, and Class Formation in Egypt, 1899–1914." *Poetics Today* 15, no. 2 (Summer 1994): 157–191.

Low, Sidney. *Egypt in Transition*. London: John Murray, 1917.

Maksudyan, Nazan. *Orphans and Destitute Children in the Late Ottoman Empire*. Syracuse: Syracuse University Press, 2014.

Manela, Erez. "Imagining Woodrow Wilson in Asia: Dreams of East-West Harmony and the Revolt against Empire in 1919." *American Historical Review* 111, no. 5 (December 2006): 1–39.

Marsot, Afaf Lutfi al-Sayyid. *Egypt and Cromer: A Study in Anglo-Egyptian Relations*. New York: Praeger, 1969.

Massey, Doreen. *For Space*. London: Sage, 2015.

McLuhan, Marshall. *The Gutenberg Galaxy: The Making of Typographic Man*. Toronto: University of Toronto Press: 1962.

———. *Understanding Media: The Extensions of Man*. New York: McGraw-Hill, 1964.

Mehrez, Samia. "Representing the Moulid: Salah Jahin's *Al-Layla al-Kabira* between Populist and Nationalist Aspirations." In *Medieval and Early Modern Performance in the Eastern Mediterranean*, edited by Arzu Öztürkmen and Evelyn Birge Vitz, 453–463. Turnhout: Brepols, 2014.

Meiton, Fredrik. *Electrical Palestine: Capital and Technology from Empire to Nation*. Berkeley: University of California Press, 2019.

El-Messiri, Sawsan. *Ibn al-Balad: A Concept of Egyptian Identity*. Leiden: Brill, 1978.

Mestyan, Adam. *Arab Patriotism: The Ideology and Culture of Power in Late Ottoman Egypt*. Princeton: Princeton University Press, 2017.

———. "Power and Music in Cairo: Azbakiyya," *Urban History* 40, no. 4 (2013): 681–704.

———. "Sound, Military Music, and Opera in Egypt during the Rule of Mehmet Ali Pasha (r. 1805–1848)." In *Ottoman Empire and European Theatre*. Vol. 2 of *The Time of Joseph Haydn—From Sultan Mahmud I to Mahmud II (r. 1730–1839)*, edited by Michael Huttler and Hans Ernst Weidinger, 631–656. Vienna: Hollitzer Wissenschaftsverlag, 2014.

———. "Upgrade? Power and Sound during Ramadan and *'Id al-Fitr* in the Nineteenth-Century Ottoman Arab Provinces." *Comparative Studies of South Asia, Africa and the Middle East* 37, no. 2 (2017): 262–279.
Mikhail, Alan. *The Animal in Ottoman Egypt*. Oxford: Oxford University Press, 2014.
Milner, Alfred. *England in Egypt*. London: Edward Arnold, 1907.
Minkin, Shane. "History from Six-feet Below: Death Studies and the Field of Modern Middle East History." *History Compass* 11, no. 8 (2013): 632–646.
Mitchell, Timothy. *Colonising Egypt*. Berkeley: University of California Press, 1991.
Moore, James. "Making Cairo Modern? Innovation, Urban Form and the Development of Suburbia, 1880–1922," *Urban History* 41, no. 1 (February 2014): 81–104.
Morrison, Heidi. *Childhood and Colonial Modernity in Egypt*. New York: Palgrave Macmillan, 2015.
al-Muwaylihi, Muhammad. *What 'Isa ibn Hisham Told Us, or, a Period of Time*. Translated and edited by Roger Allen. New York: New York University Press, 2015 [1907].
Nieuwkerk, Karin van. *Performing Piety: Singers and Actors in Egypt's Islamic Revival*. Austin: University of Texas Press, 2013.
———. *A Trade like Any Other: Female Singers and Dancers in Egypt*. Austin: University of Texas Press, 1995.
Nye, David. *Electrifying America: Social Meanings of a New Technology*. Cambridge, MA: MIT Press, 1992.
Omar, Hussein. "'Snatched by Destiny's Hand': Obituaries and the Making of Class in Modern Egypt." *History Compass* 15, no. 6 (2017): 1–12.
Pinch, Trevor and Karin Bijsterveld, eds. *Oxford Handbook of Sound Studies*. Oxford: University Press, 2011.
Pollard, Lisa. "The Family Politics of Colonizing and Liberating Egypt, 1882–1919." *Social Politics* 7, no. 1 (Spring 2000): 47–79.
———. "The Habits and Customs of Modernity: Egyptians in Europe and the Geography of Nineteenth-Century Nationalism." *Arab Studies Journal* 7–8, no. 2 (Fall 1999/Fall 2000): 52–74.
———. *Nurturing the Nation: The Family Politics of Modernizing, Colonizing, and Liberating Egypt, 1805–1923*. Berkeley: University of California Press, 2005.
Powell, Eve M. Troutt. *A Different Shade of Colonialism: Egypt, Great Britain, and the Mastery of the Sudan*. Berkeley: University of California Press, 2003.
Prestel, Joseph Ben. *Emotional Cities: Debates on Urban Change in Berlin and Cairo, 1860–1910*. Oxford: Oxford University Press, 2017.
Racy, Ali Jihad. *Making Music in the Arab World: The Culture and Artistry of Tarab*. Cambridge: Cambridge University Press, 2003.
———. "Musical Change and Commercial Recording in Egypt, 1904–1932." PhD diss., University of Illinois at Urbana-Champaign, 1977.
———. "The Record Industry and Egyptian Traditional Music: 1904–1932." *Ethnomusicology* 20, no. 1 (January 1976): 23–48.

Rath, Richard C. "Hearing American History." *Journal of American History* 95, no. 2 (2008): 417–431.
Raymond, André. *Cairo: City of History*. Cairo: American University in Cairo Press, 2007.
Reid, Donald M. *Cairo University and the Making of Modern Egypt*. New York: Cambridge University Press, 1990.
Reynolds, Bryan, and Joseph Fitzpatrick, "The Transversality of Michel de Certeau: Foucault's Panoptic Discourse and the Cartographic Impulse." *Diacritics* 29, no. 3 (1999): 63–80.
Reynolds, Nancy. *A City Consumed: Urban Commerce, the Cairo Fire, and the Politics of Decolonization in Egypt*. Stanford: Stanford University Press, 2012.
Rodaway, Paul. *Sensuous Geographies: Body, Sense and Place*. London: Routledge, 1994.
Ruiz, Mario M. "Manly Spectacles and Imperial Soldiers in Wartime Egypt, 1914–19," *Middle Eastern Studies* 45, no. 3 (2009): 351–371.
Russell, Mona. *Creating the New Egyptian Woman: Consumerism, Education, and National Identity, 1863–1922*. New York: Palgrave Macmillan, 2004.
Russolo, Luigi. *The Art of Noises: Futurist Manifestos, 1913*. New York: Something Else Press, 1967 [1913].
Ryan, James. "'Unveiling' the Tramway: The Intimate Public Sphere in Late Ottoman and Republican Istanbul." *Journal of Urban History* 44, no. 5 (2016): 1–24.
Ryzova, Lucie. *The Age of the Efendiyya: Passages to Modernity in National Colonial Egypt*. Oxford: Oxford University Press, 2014.
Sadgrove, P. C. *The Egyptian Theater in the Nineteenth Century: 1799–1882*. Reading: Ithaca Press, 1996.
Safran, Nadav. *Egypt in Search of Political Community: An Analysis of the Intellectual and Political Evolution of Egypt, 1804–1952*. Cambridge, MA: Harvard University Press, 1961.
Schafer, R. Murray. *The Soundscape: Our Sonic Environment and the Tuning of the World*. Rochester, NY: Destiny Books, 1994 [1977].
Schielke, Samuli. "Habitus of the Authentic, Order of the Rational: Contesting Saints' Festivals in Contemporary Egypt." *Critique: Critical Middle Eastern Studies* 12, no. 2 (2003): 155–172.
——. "Hegemonic Encounters: Criticism of Saints-Day Festivals and the Formation of Modern Islam in Late 19th and Early 20th-Century Egypt." *Die Welt des Islams* 47, no. 3/4 (2007): 319–355.
——. *The Perils of Joy: Contesting Mulid Festivals in Contemporary Egypt*. Syracuse: Syracuse University Press, 2012.
Schivelbusch, Wolfgang. *Disenchanted Night: The Industrialization of Light in the Nineteenth Century*. Berkeley: University of California Press, 1995.
Schmidt, Leigh Eric. *Hearing Things: Religion, Illusion, and the American Enlightenment*. Cambridge, MA: Harvard University Press, 2002.
Schölch, Alexander. *Egypt for the Egyptians: The Social-Political Crisis in Egypt, 1878–1882*. London: Ithaca Press, 1981.

Schwartz, Hillel. *Making Noise: From Babel to the Big Bang and Beyond.* New York: Zone Books, 2016.

Scott, James C. *Domination and the Arts of Resistance: Hidden Transcripts.* New Haven: Yale University Press, 1990.

Sedra, Paul. *From Mission to Modernity: Evangelicals, Reformers and Education in Nineteenth Century Egypt.* New York: I. B. Tauris, 2011.

Selim, Samah. *The Novel and the Rural Imaginary in Egypt, 1880–1985.* New York: Routledge, 2004.

Shafik, Viola. *Arab Cinema: History and Cultural Identity.* Cairo: American University in Cairo Press, 1998.

Shakry, Omnia, *The Great Social Laboratory: Subjects of Knowledge in Colonial and Postcolonial Egypt.* Stanford: Stanford University Press, 2007.

Shamir, Ronen. *Current Flow: The Electrification of Palestine.* Stanford: Stanford University Press, 2013.

Siamdoust, Nahid. *Soundtrack of the Revolution: The Politics of Music in Iran.* Stanford: Stanford University Press, 2017.

Silverstein, Shayna. "Syria's Radical Dabka." *Middle East Report,* no. 263 (2012): 33–37.

Simon, Andrew. "Censuring Sound: Tapes, Taste, and the Creation of Egyptian Culture." *International Journal of Middle East Studies* 51, no. 2 (2019): 233–256.

———. "Sounding History: Cassettes, Culture, and Everyday Life in Modern Egypt." PhD diss., Cornell University, 2017.

Smith, Anthony D. *The Ethnic Origins of Nations.* Oxford: Blackwell, 1986.

———. "Gastronomy or Geology? The Role of Nationalism in the Reconstruction of Nations." *Nations and Nationalism* 1, no. 1 (1994): 3–23.

Smith, Bruce. "Listening to the Wild Blue Yonder: The Challenge of Acoustic Ecology." In *Hearing Cultures: Essays on Sound, Listening and Modernity,* edited by Veit Erlmann, 21–42. New York: Berg, 2004.

Smith, Charles D. "'Cultural Constructs' and Other Fantasies: Imagined Narratives in *Imagined Communities*; Surrejoinder to Gershoni and Jankowski's 'Print Culture, Social Change, and the Process of Redefining Imagined Communities in Egypt.'" *International Journal of Middle East Studies* 31, no. 1 (1999): 95–102.

———. "Imagined Identities, Imagined Nationalisms: Print Culture and Egyptian Nationalism in Light of Recent Scholarship." *International Journal of Middle East Studies* 29, no. 4 (1997): 607–622.

———. "The Intellectual and Modernization: Definitions and Reconsiderations: The Egyptian Experience." *Comparative Studies in Society and History* 22 (1980): 513–533.

———. *Islam and the Search for Social Order in Modern Egypt: A Biography of Muhammad Husayn Haykal.* Albany: State University of New York Press, 1983.

Smith, Mark M. "The Garden in the Machine: Listening to Early American Industrialization." In *The Oxford Handbook of Sound Studies,* edited by Trevor Pinch and Karin Bijsterveld, 39–57. Oxford: Oxford University Press, 2011.

———. *How Race Is Made: Slavery Segregation, and the Senses*. Chapel Hill: University of North Carolina Press, 2006.

———. *Listening to Nineteenth-Century America*. Chapel Hill: University of North Carolina Press, 2001.

———. *Sensing the Past: Seeing, Hearing, Smelling, Tasting, and Touching in History*. Berkeley: University of California Press, 2008.

———. "When Seeing Makes Scents." *American Art* 24 (2010): 12–14.

Sonbol, Amira el-Azhary. *The New Mamluks: Egyptian Society and Modern Feudalism*. Syracuse: Syracuse University Press, 2000.

Soppelsa, Peter. "Urban Railways, Industrial Infrastructure, and the Paris Cityscape, 1870–1914." In *Trains, Culture, and Mobility: Riding the Rails*, edited by Benjamin Fraser and Steven D. Spalding, 117–144. Lanham, MD: Lexington Books, 2012.

Stanton, Andrea. *"This Is Jerusalem Calling": State Radio in Mandate Palestine*. Austin: University of Texas Press, 2014.

Starr, Deborah. *Remembering Cosmopolitan Egypt: Literature, Culture and Empire*. New York: Routledge, 2009.

———. "Sensing the City: Representations of Cairo's Harat al-Yahud." *Prooftexts* 26 (2006): 138–162.

Sterne, Jonathan. *The Audible Past: Cultural Origins of Sound Reproduction*. Durham, NC: Duke University Press, 2003.

———, ed. *The Sound Studies Reader*. New York: Routledge, 2012.

Stoever, Jennifer Lynn. *The Sonic Color Line: Race and the Cultural Politics of Listening*. New York : New York University Press, 2016.

Stokes, Martin. "'Abd al-Halim's Microphone." In *Music and the Play of Power in the Middle East, North Africa and Central Asia*, edited by Laudan Nooshin, 55–74. New York: Routledge, 2016.

Stolz, Daniel A. "Positioning the Watch Hand: 'Ulama and the Practice of Mechanical Time Keeping in Cairo, 1737–1874." *International Journal of Middle East Studies* 47, no. 3 (2015): 489–510.

Sykes, Jim. "Sound, Religion, and Public Space: Tamil Music and the Ethical Life in Singapore." *Ethnomusicology Forum* 24, no. 3 (2015): 485–513.

Takla, Nefertiti. "Murder in Alexandria: The Gender, Sexual and Class Politics of Criminality in Egypt, 1914–1921." PhD diss., University of California, Los Angeles, 2016.

Tam, Alon. "Cairo's Coffeehouses in the Late Nineteenth- and Early Twentieth-Centuries: An Urban and Socio-Political History." PhD diss., University of Pennsylvania, 2018.

Terry, Janice J. *The Wafd: Cornerstone of Egyptian Political Power*. London: Third World Center for Research and Publishing, 1982.

Thompson, Emily. *The Soundscape of Modernity: Architectural Acoustics and the Culture of Listening in America, 1900–1933*. Boston: MIT Press, 2004.

———. "Wiring the World: Acoustical Engineers and the Empire of Sound in the Motion Picture Industry, 1927–1930." In *Hearing Cultures: Essays on Sound, Listening and Modernity*, edited by Veit Erlmann, 191–209. New York: Berg, 2004.

Tignor, Robert L. *Modernization and British Colonial Rule in Egypt, 1882–1914*. Princeton: Princeton University Press, 1966.

———. *State, Private Enterprise, and Economic Change in Egypt, 1918–1952*. Princeton: Princeton University Press, 1984.

Tuan, Yi-Fu. *Space and Place: The Perspective of Experience*. Minneapolis: Minnesota University Press, 2014 [1977].

Tucker, Judith. *Women in Nineteenth Century Egypt*. Cambridge: Cambridge University Press, 1985.

Vatikiotis, P. J. *The History of Modern Egypt: From Muhammad Ali to Mubarak*. Baltimore: Johns Hopkins University Press, 1991.

Vitalis, Robert. *When Capitalists Collide: Business Conflict and the End of Empire in Egypt*. Berkeley: University of California Press, 1995.

Ward, Heather D. *Egyptian Belly Dance in Transition: The Raqs Sharqi Revolution, 1890–1930*. Jefferson, NC: McFarland, 2018.

Weber, Eugen. *Peasants into Frenchmen: The Modernization of Rural France, 1870–1914*. Stanford: Stanford University Press, 1976.

Wendell, Charles. *The Evolution of the Egyptian National Image: From Its Origins to Ahmad Lutfi al-Sayyid*. Berkeley: University of California Press, 1972.

Wickett, Elizabeth. *For the Living and the Dead: The Funerary Laments of Upper Egypt, Ancient and Modern*. New York: I. B. Taurus, 2010.

Williams, Gavin. "A Voice of the Crowd: Futurism and the Politics of Noise." *19th-Century Music* 37, no. 2 (2013): 113–129.

Winegar, Jessica. *Creative Reckonings: The Politics of Art and Culture in Contemporary Egypt*. Stanford: Stanford University Press, 2006.

Wishnitzer, Avner. "Into the Dark: Power, Light, and Nocturnal Life in Eighteenth-Century Istanbul." *International Journal of Middle East Studies* 46, no. 3 (2014): 513–531.

———. *Reading Clocks, Alla Turca: Time and Society in the Late Ottoman Empire*. Chicago: Chicago University Press, 2015.

———. "Shedding New Light: Outdoor Illumination in Late Ottoman Istanbul." In *Urban Lighting, Light Pollution, and Society*, edited by Josiane Meier, Ute Hasenöhrl, Katharina Krause, and Merle Pottharst, 66–84. New York: Routledge, 2015.

Woodall, Carole. "Sensing the City: Sound, Movement, and the Night in 1920s Istanbul." PhD diss., New York University, 2008.

Yousef, Hoda, *Composing Egypt: Reading, Writing, and the Emergence of a Modern Nation, 1870–1930*. Stanford: Stanford University Press, 2016.

Zayani, Mohamed. "Toward a Cultural Anthropology of Arab Media: Reflections on the Codification of Everyday Life." *History and Anthropology* 22, no. 1 (2012): 37–56.

Zirbel, Katherine. "Musical Discursions: Spectacle, Experience and Political Economy among Egyptian Performers in Globalizing Markets." PhD diss., University of Michigan, 1999.

Index

Abbas Bridge, 91
Abbas Hilmi II (Khedive), 191, 195, 198
Abbasiyya, 81, 90–92, 100–101, 104, 160, 196, 206–8
'Abd Al-Aziz Street, 192
Abdin District, 23, 68, 104, 116, 122
Abdin Palace, 104, 192, 198, 206, 209–11
'Abduḥ, Muhammad, 55
Acrobats, 201
'Adid lamentations, 176–80, 185, 187–88s
'Afiya (linguistic duel), 27
Air conditioning, 28, 126, 143, 150
Air raid sirens, 105
Alcohol, 130, 140, 202
Alexandria, 32–33, 48–49, 57–58, 70, 78, 83–103, 107, 113–14, 122–27, 129–32, 143–51, 163, 193–97, 209–10
Amusement park, 3–4, 17, 20, 145
Anglo-Egyptian Treaty (1936), 2, 89
Animal-drawn carts, 28–29, 83, 85, 94–95, 103, 107, 116, 118
Anklets, 19, 26, 37, 39
Anthems, 193–94, 197–99, 207, 213
Asphalt, 83–84, 94, 100
Ataba Square, 39, 39, 86, 92–92, 100–101, 107–10, 144, 147–48, 221, 231, 241
Auditory signaling, 8, 10, 17, 20, 34–36, 45, 51, 95, 99, 118, 157, 181, 183, 189, 191, 213
Aural modernity, 2, 5, 83–86
Automobiles, 2–4, 17, 19–20, 27–33, 75–76, 83–84, 88–89, 92, 94–107, 113, 118, 175, 199, 215

Azbakiyya district, 28, 43, 68, 86, 122, 133, 137–139, 142, 147–48, 198
Azbakiyya Garden, 49, 86, 104, 137–38, 197
Azhar Street, 92, 100, 242
Azhar University, 27, 39, 70, 86, 104 187, 205, 216

Bab al-Bahari Street, 136, 138
Bab al-Hadid Square, 100–1, 104, 108–13
Backgammon, 131, 133
Bakhit, Muhammad, 187, 202, 204–5
Bakhtin, Mikhail, 204
Bands, 1–3, 17, 20, 136, 157, 161–70, 172, 177, 182, 196–202, 207–8, 213–14
Bandstands, 197
Barak, On, 88, 114
Barbers, 122, 170
Barbershops, 121–23, 131
Bars, 108, 114, 129–32, 134, 138–39, 142–43, 146, 163
Bath procession, 168–70
Beggars, 19, 23, 26, 34–35, 40, 43–44, 52, 54–56, 65–71, 73–74, 79, 85, 108, 215, 218, 220, 223
Bells, 29, 33, 42, 60, 94–95, 97
Belly dancing, 140–41, 146, 165, 175, 221, 223
Bentham, Jeremy, 11
Bible reading, 177
Bicycle bells, 29, 31, 33
Bicycles, 27, 29, 31, 33, 94
Bourdieu, Pierre, 53–54

281

Index

Brass bands, 1–3, 17, 20 157, 163–66, 169, 196–200, 207, 213–14
Brass merchants, 27, 39–40
Brass saucers, 42, 46–48
Bribery, 58, 61, 130
British occupation, 68, 89, 122, 139, 165, 190, 197–199
Brothels, 44, 138
Bulaq, 81, 89, 91, 142
Bumper cars, 13
Buses, 30, 33, 49, 83, 89, 98–104, 108–10, 113, 117–19, 145, 218
Buza, 130

Cabarets, 20, 108, 133, 136–40, 143–48
Cabs, 49, 92, 95
Cafés, 3–4, 12, 14, 32, 40, 43, 50, 108, 121, 123, 128–48, 162–63, 222–23
Cafés chantants, 133, 138–42, 148
Cairo Electric and Gaz Administration (CEGA), 126
Cairo Electric Railways, 91
Cairo Public Motor Car Service, 99
Cairo Station (*Mahatit Misr*), 100, 108–13, 190, 209
Cairo Tramway Company, 86, 90, 92, 144
Cairo Zoo, 66, 104, 197–98
Call to prayer, 23, 48, 153, 216
Canes. *See* Walking canes
Capitulations, 89
Car horns, 20, 29, 81, 103–107, 121, 136, 175
Cars. *See* Automobiles
Carnival, 3, 201, 207
Carnival rides, 201
Carnivalesque, 201, 207
Carriages, 26–32, 44, 49, 83–84, 57, 92–96, 107–9, 113, 117, 220
Cartoons, 66, 71, 73, 136, 202, cartoon movies, 105–7
Carts, 27–31, 41, 46, 59, 83, 85, 92–96, 107–9, 118, 141–42, 171
Casino Opera, 146–47
Casinos, 81, 138, 142–44, 146–47
Castanets, 46, 140

Ceremonies, 20, 68, 90, 168, 170, 174, 184, 192–93, 197–98, 210–11
Chalcraft, John, 94
Children, 3, 17, 31, 35, 44–46, 49, 62, 65–71, 74–78, 96–97, 116, 220
Cinema, 4, 81, 104, 106, 142–50
Cinématographe Lumière, 147
Circuses, 201, 223
City squares (*Midans*), 20, 28, 32, 39, 49, 51–55, 60, 69, 83–87, 91–92, 100–4, 107–12, 118–19, 137, 144–48, 151, 190–93, 196–97, 205–08, 211, 216
Class distinction, 7–8, 20, 54, 107, 158, 160, 166, 166, 188, 215–19
Class formation, 7, 21, 53, 80, 160, 188, 216, 225
Classism, 7, 19, 53, 218
Classist discourses, 8, 53–55, 66, 80, 134, 186, 217, 220
Clot Bey Street, 101, 122
Coaches, 31, 85, 92, 94, 101, 108, 198–99
Cobblestone streets, 83, 94, 142
Coffeeshops. *See* Cafés
Colonial authority, 11, 13, 17, 52, 56, 58, 85, 114, 215
Comedy. *See* Humor
Consumption, 53, 121–22, 127, 140, 145
Coptic Christians, 177, 181, 187, 218
Coptic Churches, 177

Damietta, 129, 195
Dancing, 23, 49, 133, 139–40, 146–47, 158–61, 168–72, 175, 188, 201–2, 205, 207, 221, 223
De Certeau, Michel, 11–13, 19, 25, 32–34
Death: '*adid* lamentations, 176–80; funerary processions, 9, 157–62, 176–77, 181–87, 191–92, 198; *nadab* wails, 177–80; professional female mourners, 159, 176–81, 187–88; professional male mourners, 181–84; sounds of death, 158–59, 176–80, 183–84; wailing, 159–60, 166, 175–77;
Dominoes, 131–33

Driving, 26, 31, 34, 38, 44, 49, 89, 94–95, 105–7, 175
Drums, 17, 20–21, 94, 140–41, 165–68, 172, 177, 186, 196, 207–8, 213, 220–21

Early, Evelyn, 40
Ear-witnessing, 26
Effendiyya, 30, 36, 105–7
Egyptian Government, 2, 20, 57–61, 63–64, 68, 71, 77, 85–86, 94, 97, 103, 124–26, 129, 131–35, 150, 162, 190–94, 199, 204–7, 210, 214
Egyptian Government Railway, 103, 124
Eid, 46, 161, 184, 196–97, 213
Electric appliances, 4, 126–28, 152
Electricity: affordability of, 136, 150–52; domestic uses of, 126–27, 134; implications of, 4, 5, 20, 84–86, 120–26, 153, 162, 216; power plants, 122–26
Electrification, 19–20, 120–28, 137–38, 152, 215
Embodiment: and celebration, 20, 157–60, 166, 168, 175–76, 183, 193, 202, 217–18; critiques against, 19, 186, 189, 218–19; embodied grieving, 159, 176–78; embodied sounds, 2, 16, 19–21, 33–37, 52, 99–100; embodiment in the tram, 113–18; everyday embodiment, 3, 5–6, 12–13, 25–27, 32, 79, 84, 182, 215–16; and the senses, 16, 21, 79, 117
Empain, Baron Édouard Louis Joseph, 91
Ener, Mine, 66
Engine noise, 4, 28, 83, 95, 100, 215
Entertainment, 3–5, 19, 44, 49, 54, 65, 68, 71, 79, 121, 127–32, 137–153, 168–72, 188, 198, 201–2, 218, 221–22
Everyday life, 3–14, 17–21, 25–26, 31–34, 51–54, 63, 73, 79, 84, 100, 108, 119, 121, 157, 214–25
Exhibitions: the 1936 Cairo Agricultural and Industrial Exhibition, 1–3, the 1889 World Exhibition in Paris, 2

Factories, 40, 90, 114, 124, 127, 152, 223
Fahmy, Khaled, 85
Faruq I (King), 2, 59, 61, 77, 194, 209
Fear of the crowd, 19, 39, 53, 56, 66–68, 85, 158
Feasts 191, 201, 204
Fellahin, 36, 66, 89, 110–14, 158–60, 217, 219–20, 223–24
Female Quran reciters, 65–66
Festivities, 20–21, 23, 157–60, 168, 186, 189–94, 198, 201–4, 207, 213, 217
Fireworks, 60, 189–90, 201, 207, 213
Food, 7, 45, 60–61, 74–76, 108–9, 172, 184, 190, 194, 201, 223
Footsteps, 4, 19, 33–35, 38, 213
Foreigners, 18, 40, 90, 92, 97, 120, 124–25, 130, 132, 191, 199
Fortune tellers, 44, 54, 65, 69, 132, 223
Foucault, Michel, 11–12
Frenkel Productions, 106
Funerals, 9, 20, 23, 49, 66, 157–62, 175–91, 201, 214–16, 218

Galloping horses, 27, 33, 94–95
Gamal, Samia, 146
Gambling, 69, 129
Gaslight, 86, 120–24, 137, 144, 153
Ghafir. *See* Night guards
Ghuriyah Street, 162
Gender-mixing, 189–90, 114, 189, 205, 218
Giza, 86, 90–91, 104, 140 146, 198
Great Britain, 2, 68, 89, 122, 165, 192, 197, 209
Greeks, 116, 132
Groppi, 132
Guilds, 58
Gymnasts, 139
Gypsies, 66, 68–69

Hasaballah Brass Band, 163–66, 177, 214
Hashish, 131

Haussmannization, 107
Heliopolis, 90–91, 100, 104, 125, 144–45, 148
Heliopolis Oases Company, 91, 144
Henna Night, 170
Highways, 89
Hirschkind, Charles, 6
Hissing, 94
Hollywood films, 150–51
Homelessness, 54, 65–66, 68, 70–73, 75, 77
Honking horns, 27, 33, 60, 83, 89, 95, 97, 103–107, 121, 136, 175
Hospitals, 40, 70–71
Humor, 1, 105–6, 139, 174, 182, 222
Husayn District, 26–29, 39, 46, 86, 152

Imad al-Din Street, 104, 132, 138, 142, 145–49
Imbaba, 26, 48, 86, 90–91, 147
Indoor space, 4, 43, 126–28, 143–44, 150–52, 160, 166, 175
Infrastructure: electrification, 19–20, 120–28, 137–38, 152, 215; indoor plumbing, 41–43; municipal gas, 86, 120–24, 137, 144, 153; road building, 86–89, 92–94, 100–3, 118
Intersensorality, 13–14
Ismail (Khedive), 85, 100, 190–92, 199
Ismailia, 28, 85–86, 89, 127, 129–30
Italians, 89

Jingles, 19, 33, 36–39, 94
Jokes, 12, 27, 57, 110, 174
Jugglers, 139, 168
Juvenile delinquency, 70
Juvenile reformatories, 75, 77

Karioka, Tahiyya, 146
Khan al-Khalili Bazar, 27, 39, 86
Khanzindar Square, 137
Khulkhal. See Anklets
King Faruq I, 2, 59, 61, 77, 194, 209–10
King Fuad I, 2, 155, 192–93, 199–200, 209

Lamentations, 166, 176–81
Laughter, 57, 204
Laws: anti-noise, 18, 29, 103–7, 121, 134; anti-vagrancy, 68–78; traffic, 31, 101–7, 129, 136, 162
Lebon Corporation, 124–26, 144
Lefebvre, Henri, 12, 79
Leisure, 5, 20, 121, 128, 137, 153
Lemonade sellers, 31, 38, 42, 46, 49
Lightbulbs, 127, 144
Lights: arc, 123, 144–45; gas, 86, 120–24, 137, 144, 153; electric, 2, 5, 20, 120–24, 136–38, 144, 153, 162, 168, 193–94, 198, 206–7, 211, 216
Liminal spaces, 50–51, 84, 115–16
Loudness, 1, 3–5, 7–9, 12, 20, 27, 29, 33–35, 37, 39, 46, 50, 52–57, 74, 90, 94–96, 99–100, 103, 106–8, 117, 132, 136, 141–43, 152–53, 157–61, 167–69, 174, 177–78, 184–90, 193, 199, 201, 214, 217–20
Loudspeaker, 1–4, 17, 20–21, 121, 136, 150, 153, 157, 163, 166–68, 175, 184, 189–90, 208–16
Lower classes, 7–8, 35, 56, 74, 94, 107, 149, 158, 168, 189, 217–18
Lower-middle classes, 37, 94
Luna Park, 3, 145

Maadi, 90
Magicians, 10, 139, 201–2, 221
Mahmal, 190–91, 204–5, 207
Marketplace, 4, 14, 25, 40, 50, 75, 114, 118, 121, 124, 142, 151
Marsa Matruh, 89
Masabni, Badiʿa, 146–47
Mawalid, 46, 158, 161–62, 166, 189–90, 201–7, 213–14, 217, 223–24
Merchants, 19, 26–27, 34, 40–48, 51, 57–62, 64, 76, 79, 83, 88, 108, 116, 123, 169, 194, 216
Mestyan, Adam, 163
Microphones, 4, 21, 27–28, 153, 162, 168, 209–10, 216
Midan. See City squares

Middle classes, 5, 7–9, 20, 55–56, 71, 90–92, 101, 113–14, 158, 160, 166, 185, 188, 217–20
Mishmish Effendi cartoons, 105–7
Mitchell, Timothy, 11
Mitwali, Ahmad. 106
Modernist discourses, 7, 158, 185, 217–18
Mosques, 39, 65, 70, 73, 75, 86, 91, 130, 153, 187, 193, 214, 216
Mother-in-law, 174–75
Motorcycles, 4, 17, 30, 53, 83, 98, 100, 102, 199
Mourners, 159, 176–84, 187–90
Movie industry, 151
Movie theaters, 20, 50, 75, 105, 121, 138, 147–51; outdoor, 143, 152, 197
Muʻadidat, 176–80, 182
Mubarak, Ali (Pasha), 85, 130, 201
Muhammad Ali Street, 86, 91, 100, 165, 190, 205
Muʻiz Street, 26–29
Multisensory experiences, 3, 12, 14–15, 17, 21, 33, 85, 193, 201, 213
Mundane sounds, 19–21, 23–40, 45–52, 108–18, 120–21, 216–21
Municipal lights, 4, 86, 120, 122–26, 137, 144, 153
Municipal water, 41–43. 85–86
Music: state, 163, 193–94, 196–201, 207, 213–14; street, 17, 20–21, 94, 140–41, 157, 163–68, 172, 177, 186, 196, 207–8, 213–14, 220–21; wedding, 9, 20, 23, 157–75, 180, 185–88, 214–18
Muski Street, 38, 49, 86, 92, 100, 162

Nahas, Mustafa (Pasha), 211–12
Nahasin Street, 162
Nationalism, 2, 9, 125, 215, 217–24
Neon signs/lights, 143–46
Newspaper boys, 108, 116–17
Night guards, 34–35
Nightlife, 5, 20, 84, 120–23, 134–45, 153
Noise, 1–7, 15–18, 34, 120, 123, 128, 133, 160, 190, 213–17; and class, 7–9, 74, 134, 189, 204, 217–20; and radios, 121, 136–37; and traffic, 20, 26–31, 97, 103–13, 128
Noise abatement, 18, 29, 103–7, 121, 134
Non-verbal vocalizations, 20, 35, 60, 94, 103, 157–60, 178
Nostalgia and the senses, 17, 120, 163, 215, 220–24

Ocularcentrism, 9–11
Olfactory sense, 14–16, 32, 85, 181
Opera Square, 49, 104, 146–48, 190–92
Orality, 9–10
Ordinary people, 3, 11–14, 18–21, 25–26, 51–52, 54, 79, 84, 128, 155–58, 191, 194–96, 205, 214–18, 222–25
Orphanages, 70–71, 77–78
Orphans, 70–78

Panopticon, 11–12, 218
Parades, 19, 21, 23, 163, 168, 175, 190–92, 196–98, 204–8, 213–15, 220
Passersby, 27, 31, 42, 56, 131–32, 134, 144
Paving streets, 27, 83–85, 88–89, 94–100, 129
Peasants. See *Fellahin*
Peddlers. *See* Street hawkers
Pedestrians, 3, 19, 25–35, 51, 83, 94–98, 103–6, 116, 119, 132, 141, 215, 218
Piety, 184, 188, 200–1, 213
Petitions, 18, 58, 61–64, 122–23, 135, 195, 202
Pickpockets, 56, 66, 115–17, 217–18
Pilgrimage procession, 161, 168, 170, 190–91, 204
Police, 17, 23, 32, 54, 57, 68–70, 77–78, 101, 103, 107, 134, 163–66, 190, 192, 195, 198; abuse, 57–65, 69–73; policing the streets, 54, 57–60, 79–81; traffic, 103–6;
Politics of sound, 3, 5, 217–20
Popular culture, 165, 174
Popular Islam, 8, 201–8, 213–14
Poverty, 35, 63–65, 70–74, 220

Powerplants, 122–26
Prince Faruq Road, 92, 100, 194, 209
Print media, 8, 10
Private space, 41, 50–51, 84, 115–17
Processions, 9, 19–20, 23, 157–77, 181–84, 192–97, 220–21
Professional mourners: female, 159, 176–81, 184, 187–88; male, 181–83
Propaganda, 4, 20, 208, 211, 216
Prostitution, 44, 56, 69, 138
Public hygiene, 60, 85
Public order, 19, 54, 63, 66, 85, 189
Public space, 13, 23, 79, 114–19, 135, 189
Public transportation, 83, 101–3, 114–17
Puppet shows, 201, 223

Qasr al-Nil Street, 91, 104
Quran recitation: 162, 177, 182–84, 187, 194, 208, 211, 213; female reciters, 65–66

Radio, 1–2, 126–28, 134–36, 152–53, 175, 199, 218, 223; noise of, 121, 134–37, 152–53; speeches broadcast on loudspeakers, 21, 208–11; 213–14
Radwan, Fathi, 23, 45–46, 120, 131, 143, 150, 163, 178, 182–83, 221
Ramadan, 23, 46, 190–91, 194–97, 213, 220
Ramadan cannons, 194–97
Raml District (Alexandria), 90, 151
Rasdikhana Square, 196
Rath, Richard Cullen, 15
Red light districts, 137–38
Reformatories, 70, 75–77
Refrigerators, 4, 127–28, 152
Restaurants, 122–23, 130, 144–45
Revolution: the 1919 Revolution, 108, 132; the 2011 Revolution, 108, 188
Rihani, Naguib, 132, 137, 143, 146
Roads. *See* Streets
Roda Island, 91
Royal anthems, 193–94, 197–99
Royal authority, 73, 155–57, 189–94, 204, 207, 213

Rud al-Farag District, 75, 91, 104, 141–43
Rumors, 4, 12, 17–18
Rural areas, 35, 40–41, 220
Rural roads, 34, 88
Rush-hour, 33–34, 108, 118

Saint's Day. See *Mawalid*
Satirical press, 99–100, 193, 202
Sayyida Zaynab District, 45–46, 65, 75–76, 91, 120, 163, 178, 182, 220
Schafer, R. Murray, 15, 26, 44, 189, 204, 222
Schielke, Samuli, 201
Schmidt, Leigh Eric, 9
Screams, 66, 74, 159, 176–78, 180
Sensorium, 9, 30–31, 83–84, 118–19, 194, 215, 224
Sensory marginalization, 7–8, 52, 80, 190, 201, 217–20, 224
Sexuality, 57, 173–74
Shawadir [canopy tents], 160–63, 167–68, 184, 193
Shibin al-Kum, 193, 263
Shobra, 90–91
Shouting, 12, 27, 46, 50, 117, 133, 178–79
Shrieking 159, 168, 175–76, 183, 218
Sidewalks, 25–28, 31, 50, 75, 95, 103, 129, 132, 162, 218
Silencing street cries, 54, 60–68
Silencing the poor, 7–9, 19, 54, 65–80, 158, 213–14, 217–20
Silent movies, 140, 150
Smells, 3–4, 8, 13–18, 25, 32, 83–85, 99–100, 118–19, 182, 187, 217, 221
Smith, Mark, 5, 10
Société Anonyme des Tramways du Caire, 90–91
Société des Automobiles et Omnibus du Caire, 99
Société Générale des Chemins de Fer Économiques, 90
Sounds: of automobiles 2–4, 17, 19–20, 27–33, 75–76, 83–84, 88–89, 92,

94–107, 113, 118, 175, 199, 215; of carriages and carts, 26–32, 41, 44, 46, 49, 57–59, 83–85, 92–96, 107–9, 113, 117–18, 141–42, 171, 220; of celebration, 157–76, 193–94, 199, 204–8, 213–26; of footsteps, 4, 19, 33–35, 38, 213; of jewelry, 19, 26, 36–39; of street hawkers' calls, 18–19, 26, 31, 35, 40, 43–68, 79, 85, 108–9, 113, 116, 118, 121, 132, 216–18, 223–24
Soundscapes, 1–21, 26, 30–33, 43–47, 83–84, 103, 118–20, 136, 147, 153, 157–58, 166, 177, 198, 216, 222–25
Spectacles, 19–20, 38, 155, 157–58, 191–92, 213–14
Speeches, 16, 18, 21, 153–55, 162, 194, 208–19
Squares. *See* City squares
State music, 163, 193–94, 196–201, 207, 213–14
Station Square (Cairo), 100–4, 109–12
Sterne, Jonathan, 6
Stoever, Jennifer, 7
Storytellers, 132, 135
Street children, 35, 44, 65–70, 74–78
Street hawkers, 3, 18–19, 26, 31, 35, 40, 43–68, 79, 85, 108–9, 113, 116, 118, 121, 132, 216–18, 223–24
Street life, 3, 5–6, 13–14, 19–21, 25, 54, 119, 216, 220–23,
Streetlights, 120–24, 133, 152
Street music, 20, 157, 163–75, 214
Street sounds, 5, 7, 13, 19–21, 23, 26–38, 43–49, 50–52, 60–62, 189–90, 213–15, 220–25
Streets: asphalt/paved, 27, 83–89, 94, 100, 129; cobblestone, 83, 94, 142; dirt, 88, 95, 162; macadamized, 88, 101, 129, 162
Sufis, 71–73, 99, 201–3, 207–8
Syrians, 132

Takht, 142, 172
Talkies, 147–52

Tanta, 73, 129, 195, 209
Taste, 7–8, 14–16, 53–54, 162, 166, 187, 201, 219
Taxis, 49, 98, 101–2, 108–10
Technology, 4–10, 12, 15, 26, 33, 51, 81–84, 94, 113, 118–21, 123–28, 134, 136, 147, 151, 189, 215–18
Telephones, 1, 4, 10, 84, 120, 152, 176, 210
Theaters, 132, 143–49, 197, 201
Theft, 56, 66, 74–75, 115–17, 217–18
Thompson, Emily, 147
Touch, 13–16, 25, 31, 100, 117–18
Train stations, 91, 99, 104, 109–13, 193, 209
Trains, 1–3, 83, 98–99, 107–13, 209
Traffic, 9, 17, 19–20, 25–33, 38, 49, 51, 60–63, 81, 83–89, 90–95, 98–109, 115–19, 129, 136, 152, 162, 214–20
Traffic laws, 31, 101–107, 129, 136, 162
Tram surfing, 115–118
Tramways, 33, 39, 84–88, 90–97, 101, 104, 108–10, 113–18
Transportation hubs, 20, 86, 91–92, 107–13, 119, 153
Tuan, Yi-Fu, 13

Ululations, 20, 157–60, 166, 168, 175–76, 183, 193, 217–18
'Umar, Muhammad, 44–45, 54–57, 65–66, 74, 186
Undertakers, 181–83
Unions, 61–63
Upper classes, 9, 21, 36, 49, 56, 75, 90, 101, 113, 160, 173, 175, 218, 224
Upper Egypt, 66, 110, 143, 166, 172, 179–81, 194–96, 223
Upper-middle classes, 21, 90, 114, 122, 160, 218
Upward mobility, 8, 36, 53, 55, 71, 80, 127, 160, 217–18
Urban soundscapes, 4, 31, 43, 158, 222
Urbanization, 5, 7, 83, 113, 160, 221
Utilities, 20, 120, 124–25, 145
Utility companies, 70–74, 88, 124–25

Vagrancy laws, 68–78
Vendors, 107, 108, 132, 136, 141–42, 182, 215, 218, 221
Verbal vocalizations, 20, 60, 106, 157–59, 169, 178
Vitalis, Robert, 125
Vocalizations, 20, 34, 94, 157–60
Voices, 1, 7, 18, 21, 25, 40–44, 47–48, 50, 66, 95, 107, 118, 147, 184–88, 213, 221–22, 225
Vulgarization, 8, 53, 56, 166, 217–19, 221, 224

Wafd Party, 9, 211
Wails, 20, 157, 160, 176–78, 218
Wajh al-Birkah Street, 137–38
Walking canes, 34, 36–37, 84
Walking the city, 3, 12–13, 19, 25–26, 31–42, 60, 106, 114, 182, 187
Water companies, 41, 88
Water plumbing, 41–43
Water pumps, 3–4, 83
Water sellers, 26, 41–44, 46, 49, 107, 222
Weddings, 9, 20, 23, 157–75, 185–88, 214–18
Whistling, 1, 29
Wickett, Elizabeth, 179

World War I, 91, 122–23, 126, 138–39, 141, 144, 144, 149, 233
World War II, 63, 89, 105, 127, 136, 150, 168, 216
Women: female street hawkers/vendors, 37, 44–46, 48, 51–52, 56–57, 66; female merchants, 38–40; female water carriers, 42–43; female Quran reciters, 65–66; female mourners, 159, 176–84, 187–90; female pedestrians, 36–38
Wedding songs, 172–75, 180
Working classes, 7, 36, 90, 113–14, 121, 135–36, 152, 158, 160, 164, 186, 189, 213, 218–20, 223–24

Yelling, 27, 30, 39, 45, 60, 94–106, 202
Young Egypt, 143, 221

Zaffah, 161, 165–66, 168–75, 186, 221
Zagharit. *See* Ululations
Zaghlul, Ahmad Fathi, 55
Zaghlul, Saad, 55, 155
Zamalek, 2, 86, 90–91, 104
Zar, 158, 189, 267
Ziftah, 129, 195–96
Zoo, 66, 104, 197–98

The authorized representative in the EU for product safety and compliance is:
Mare Nostrum Group
B.V Doelen 72
4831 GR Breda
The Netherlands

www.ingramcontent.com/pod-product-compliance
Lightning Source LLC
Chambersburg PA
CBHW031759220426
43662CB00007B/457